THE WAY OF THE WILD

F. ST. MARS

ZINC READ

Published by
This edition copyright © 2024 by Zinc Read
Design and setting By Zinc Read
Email- zincread@gmail.com
ISBN-978-93-6277-402-6

CONTENTS

I 1
II 18
III 30
IV 44
V 56
VI 68
VII 82
VIII 91
IX 101
X 113
XI 119
XII 134
XIII 144
XIV 158
XV 167
XVI 179
XVII 194
XVIII 208
THE END 219

I

GULO THE INDOMITABLE

If his father had been a brown bear and his mother a badger, the result in outward appearance would have been Gulo, or something very much like him. But not all the crossing in the world could have accounted for his character; that came straight from the Devil, his master. Gulo, however, was not a cross. He was himself, Gulo, the wolverine, *alias* glutton, *alias* carcajou, *alias* quick-hatch, *alias* fjeldfras in the vernacular, or, officially, *Gulo luscus*. But, by whatever name you called him, he did not smell sweet; and his character, too, was of a bad odor. A great man once said that he was like a bear cub with a superadded tail; but that great man cannot have seen his face. If he had, he would have looked for his double among the fiends on the top of Notre Dame. There was, in fact, nothing like him on this earth, only in a very hot place not on the earth.

He was, in short, a beast with brains that only man, and no beast, ought to be trusted with; and he had no soul. God alone knows if love, which softens most creatures, had ever come to Gulo; his behavior seemed to show that it had not. Perhaps love was afraid of him. And, upon my soul, I don't wonder.

It was not, however, a hot, but a very cold, place in the pine-forest where Gulo stood, and the unpitying moon cast a dainty tracery through the tasseled roof upon the new and glistening snow around him--the snow that comes early to those parts--and the north-east wind cut like several razors. But Gulo did not seem to care. Wrapped up in his ragged, long, untidy, uncleanly-looking, brown-black cloak--just his gray-sided, black fiend's face poking out--he seemed warm enough. When he lifted one paw to scratch, one saw that the murderous, scraping, long claws of him were nearly white; and as he set his lips in a devilish grin, his fangs glistened white in the moonlight, too.

Verily, this was no beast--he would have taped four feet and a quarter from tip to tip, if you had worn chain-mail and dared to measure him--no beast, I say, to handle with white-kid ball gloves. Things were possible from him, one felt, that were not possible of any other living creature--awful things.

Suddenly he looked up. The branches above him had stirred uneasily, as if an army were asleep there. And an army was--of wood-pigeons. Thousands upon thousands of wood-pigeons were asleep above his head, come from Heaven knows where, going to--who could tell in the end?

All at once one fell. Without apparent reason or cause, it fell. And the wolverine, with his quick, intelligent eyes, watched it fall, from branch to branch, turning over and over--oh! so softly--to the ground. When he had poked his way to it--walking flat-footed, like a bear or a railway porter--it was dead. Slain in a breath! Without a flutter, killed! By what? By disease--diphtheria. But not here would the terrible drama be worked out. This was but an isolated victim, first of the thousands that would presently succumb to the fell disease far, far over there, to the westward, hundreds of miles away, in England and Wales, perhaps, whither they were probably bound.

But the poor starved corpse, choked to death in the end maybe, was of no use to the wolverine. As he sniffed it he found that out. The thing was wasted to the bones even. And turning away from it--he suddenly "froze" in his tracks where he stood.

One of those little wandering eddies which seem to meander about a forest in an aimless sort of way, coming from and going now hither, as if the breeze itself were lost among the still aisles, had touched his wet muzzle; and its touch spelt--"Man!"

If it had been the taint of ten thousand deaths it could not have affected him more. He became a beast cast in old, old bronze, and as hard as bronze; and when he moved, it was stiffly, and all bristly, and on end.

Animals have no counting of time. In the wild, things happen as swiftly as a flash of light; or, perhaps, nothing happens at all for a night, or a day, or half a week. Therefore I do not know exactly how long that wolverine was encircling that scent, and pinning it down to a certain spot--himself unseen. All animals, almost, can do that, but none, not even the lynx or the wild cat, so well as the wolverine. He is the one mammal that, in the wild, is a name only--a name to conjure with.

He found, in the end, that there was no man; but there *had* been. He found--showing himself again now--that a man--a hunter, a trapper, one after fur--had made himself here a *cache,* a store under the earth; and--well, the wolverine's great, bear-like claws seemed made for digging.

He dug--and, be sure, if there had been any danger there he would have known it. He dug like a North-Country miner, with swiftness and precision, stopping every now and again to sit back on his haunches, and, with humped shoulders, stare--scowl, I mean--round in his lowering, low-browed fashion.

Once a bull-elk, nearly a six-footer, but he loomed large as an elephant, came clacking past between the ranked tree-boles, stopping a moment to straddle a sapling and browse; while the wolverine, sitting motionless and wide-legged, watched him. Once a lynx, with its eternal, set grin, floated by, half-seen, half-guessed, as if a wisp of wood mist had broken loose and was floating about. Once a fox, somewhere in the utter silence of the forest depths, barked a hoarse, sharp, malicious sound; and once, hoarser still and very hollowly, a great horned owl hooted with disconcerting suddenness. (The scream of a rabbit followed these two, but whether fox or owl had been in at that killing the wolverine never knew.) Twice a wood-hare turning now to match the whiteness of its surroundings, finicked up one of the still, silent forest lanes towards him, stopped, faced half-round, sat "frozen" for a fraction, and vanished as if it were a puff of wind-caught snow. (And, really, one had no idea till now that the always apparently lifeless forest could have been so full of life in the dark hours.)

But all these things made no difference to the wolverine, to Gulo, though he "froze" with habitual care to watch them--for your wild creature rarely takes chances. Details must never be overlooked in the wild. He dug on, and in digging came right to the *cache,* roofed and anchored all down, safe beyond any invasion, with tree-trunks. And--and, mark you, not being able to pull tree-trunks out of the ground, and being too large to squeeze between them, he gnawed through one! Gnawed through it, he did, and came down to the bazaar below.

So far, he had been only beast. Now we see why I said he had more brains than were good for any animal except man.

He bit through the canvas, or whatever it was that protected the *cached* articles. He got his head inside. He felt about purposefully, and backed out, dragging a trap with him. With it he removed into the inky shadows, and it was never found again.

He returned. He thrust his head in a second time, got hold of something, and backed out. It was another trap, and with it he vanished also; and it, too, was never found. He returned, and went, and a third trap went with him.

The fourth investigation revealed an ax. It he partly buried. The fifth yielded a bag of flour, which he tore up and scattered all over the place. The sixth inroad produced a haunch of venison, off which he dined. The seventh showed another haunch, and this he buried somewhere unseen in the shades. The eighth overhaul gave up some rope, in which he nearly got himself entangled, and which he finally carried away, bitten and frayed past use. The ninth search rewarded him with tea, which he scattered, and bacon, which he buried.

What he could not drag out, he scattered. What he failed to remove, he defiled. And, at last, when he had made of the place, not an orderly *cache*, but a third-rate *debacle*, he sauntered, always slouching, always grossly untidy, hump-backed, stooping, low-headed, and droop-tailed, shabbily unrespectable, out into the night, and the darkness of the night, under the trees.

By the time day dawned he was as if he never had been--a memory, no more. Heaven knows where he was!

Gulo appeared quite suddenly and very early, for him, next afternoon, beside some tangled brush on the edge of a clearing. He was sitting up, almost bolt-upright, and he was shading his eyes with his forepaws. A man could not have done more. And, in fact, he did not look like an animal at all, but like some diabolically uncouth dwarf of the woods.

A squirrel was telling him, from a branch near by, just what everybody thought of his disgraceful appearance; and two willow-grouse were clucking at him from some hazel-tops; whilst a raven, black as coal against the white of the woods, jabbed in gruff and very rude remarks from time to time.

But Gulo was taking no notice of them. He was used to attentions of that kind; it was a little compliment--of hate--they all paid him. He was looking persistently down the ranked, narrowing perspective of the buttressed forest glade to where it faded in the blue-gray mist, southward, as if he expected something to come from there. Something was coming from there now; and there had been a faint, uneasy sort of whisper in that direction for some time. Now it was unmistakable.

A cow-elk, first of the wary ones to move on alarm, came trotting by, her Roman nose held well out; a red-deer hind, galloping lightly like some gigantic hare, her big ears turned astern; a wolf, head up, hackles alift, alternately loping and pivoting, to listen and look back, a wild reindeer, trotting heavily, but far more quickly than he seemed to be--all these passed, now on one side, now on the other, often only glimpses between the tree-boles, while the wolverine sat up and shaded his eyes with his paws. Something was moving those beasts, those haunters of the forest, and no little thing either. Something? What?

Very softly down the glade runs a waiting, watching shade,
And the whisper spreads and widens far and near;
And the sweat is on thy brow, for he passes even now--
He is Fear, O Little Hunter, he is Fear.

Down came Gulo in that grim silence which was, except for his domestic arguments, characteristic of the beast, and trotted to a pool hard by. The pool was spring-fed, and covered, as to every dead leaf and stone, with fine green moss of incomparable softness. He drank swiftly and long, then flung about with a half-insolent, half-aggressive wave of his tail, and set off at a rolling, clumsy, shuffling shamble.

At ordinary times that deceiving gait would have left nearly everything behind, but this afternoon it was different. Gulo had barely shed the shelter of the dotted thickets before he realized, and one saw, the fact. He broke his trot. He began to plunge. Nevertheless, he got along. There was pace, of a sort. Certainly there was much effort. He would have outdistanced you or me easily in no time, but it was not you or I that came, and who could tell how fast that something might travel?

The trouble was the snow--that was the rub, and a very big and serious rub, too, for him. Now, if the snow had been a little less it would not have mattered--a little more, and he could have run easily along the hard crust of it; but it was as it was, only about two feet, just enough to retard him, and no more. And it is then, when the snow is like that, just above a couple of feet deep, that man can overtake friend wolverine--if he knows the way. Most men don't. On that he trusted. At any other time--but this was not any other time.

Sound carries a long way in those still parts, and as he hurried Gulo heard, far, far behind in the forest, the faint, distant whir of a cock-capercailzie--the feathered giant of the woods--rising. It was only a whisper, almost indistinguishable to our ears, but enough, quite enough, for him. Taken in conjunction with the mysterious shifting of the elk and the red deer and the reindeer and the wolf, it was more than enough. He increased his pace, and for the first time fear shone in his eyes--it was for the first time, too, in his life, I think.

A lynx passed him, bounding along on enormous, furry legs. It looked all legs, and as it turned its grinning countenance to look at him he cursed it fluently, with a sudden savage growl, envious, perhaps, of its long, springing hindlegs. Something, too--the same something--must have moved the lynx, and Gulo shifted the faster for the knowledge.

Half-an-hour passed, an hour slid by, and all the time Gulo kicked the miles behind him, with that dogged persistency that was part of his character. Nothing had passed him for quite a while, and he was all alone in the utterly still, silent forest and the snow, pad-pad-padding along like a moving, squat machine rather than a beast.

At last he stopped, and, spinning round, sat up. A gray-blue haze, like the color on a wood-pigeon, was creeping over everything, except in the west, where the sky held a faint, luminous, pinky tinge that foretold frost. It was very cold, and the snow, which had never quite left off, was falling now only in single, big, wandering flakes. The silence was almost terrifying.

Then, as Gulo sat up, from far away, but not quite so far away, his rounded ears, almost buried in fur, caught faintly--very, very faintly--a sound that brought him down on all fours, and sent him away again at a gallop with a strange new light burning in his little, wide-set eyes. It was the unmistakable sound of a horse sneezing--once. Gulo did not wait to hear if it sneezed twice. He was gone in an instant. Man, it seemed, had not been long in answering that challenge of the *cache* escapade.

After that there was no such thing as time at all, only an everlasting succession of iron-hard tree-trunks sliding by, and shadows--they ran when they saw him, some of them, or gathered to stare with eyes that glinted--dancing past. The moon came and hung itself up in the heavens, mocking him with a pitiless, stark glare. (He would have given his right forepaw for a black night and a blinding snowstorm.) It almost seemed as if they were all laughing at him, Gulo the dreaded, the hated hater, because it was his turn at last, who had so freely dealt in it, to know fear.

Hours passed certainly, hours upon hours, and still, his breath coming quickly and less easily now with every mile, Gulo stuck to the job of putting the landscape behind him with that grim pertinacity of his that was almost fine.

At last the trees stopped abruptly, and he was heading, straighter than crows fly, across a plain. The plain undulated a little, like a sea, a dead sea, of spotless white, with nothing alive upon it--only his hunched, slouching, untidy, squat form and his shadow, "pacing" him. At the top of the highest undulation he stopped, and glowered back along the trail.

Ahead, the forest, starting again, showed as a black band a quarter of an inch high. Behind, the forest he had already left lay dwarfed in a ruled, serried line. But that was not all. Something was moving out upon the spotless plain of snow, something which appeared to be no more than crawling, ant-like, but was really traveling very fast. It looked like a smudged dot, nothing more; but it was a horse, really, galloping hard, with a light sleigh, and a man in it, behind. The horse had no bells, and it was not a reindeer as usual. Pace was wanted here, and the snow was not deep enough to impede the horse, who possessed the required speed under such conditions.

The horse had been trotting along the trail, till it came to the place where Gulo had looked back and heard the sneeze, and knew he was being followed. Then it had started to gallop, and, with ears back and teeth showing, had never ceased to gallop. This, apparently, was not the first wolverine that horse had trailed. It seemed to have a personal grudge against the whole fell clan of wolverine, and to be bent upon trampling Gulo to death.

Gulo watched it for about one quarter of a second. Then he quitted, and the speed he had put up previously was nothing to that which he showed now--uselessly. And, far behind him, the man in the sleigh drew out his rifle from under the fur rugs. He judged that the time had about come. The end was very near.

But he judged wrong. Gulo made the wood at length. With eyes of dull red, and breath coming in short, rending sobs, he got in among the trees. He did it, though the feat seemed impossible, for the trees had been so very far away. Got in among the trees--yes, but dead-beat, and--to what end? To be "treed" ignominiously and calmly shot down, picked off like a squirrel on a larch-pole. That was all. And that was the orthodox end, the end the man took for granted.

In a few minutes the horse was in the forest too, was close behind Gulo. In spite of the muffling effect of snow, his expectant ears could hear the quadruple thud of the galloping hoofs, and--

Hup! Whuff! Biff-biff! Grrrrrr! Grr-ur-ururrh! Grrrr-urr!

It had all happened quick as a flash of light. A huge, furry, reeking mass rising right in the wolverine's path from behind a tree, towering over him, almost mountainous to his eyes, like the very shape of doom! Himself hurling sideways, and rolling over and over, snarling, to prevent the crowning disaster of collision with this terrible portent! A blow, two blows, with enormous paws whose claws gleamed like skewers, whistling half-an-inch above his ducked head! Jaws, monstrous and wet, grabbing at him in enraged confusion, and rumblings deep down in the inside of the thing that ran cold lightning-sparks all up his spine. That was what Gulo saw and heard.

The wolverine rolled, clawing and biting, three times, and without a pause sprang to his feet again, and leapt madly clear, stumbled on a hidden tree-root, rolled over again twice, and up, and hurled, literally with his last gasp and effort, headlong through the air behind a tree-bole, where he remained all asprawl and motionless, except for his heaving sides, too utterly done at last for *any* terror to move him.

There followed instantly a horse's wild snort; another; a shout; the crack of a rifle cutting the silence as a knife cuts a taut string; another crack; an awful, hoarse growl; the furious thudding of horse's hoofs stampeding and growing fainter and fainter; and an appalling series of receding, short, coughing, terrifying, grunting roars. Then silence and utter stillness only, and the cold, calm moon staring down over all.

Gulo picked himself up after a bit, and slouched round the tree to investigate. He found tracks there, and blood; and the tracks were the biggest footprints of a bear--a brown bear--that he had ever come across, and I suppose that he must have sniffed at a few in his time.

Presumably the man had fired at the bear when the startled horse shied. Presumably, too, the bear was hit. He had gone straight away in the track of horse and man, anyway, and--he had saved the wolverine's life, after, with paw and teeth, doing his best to end it. Possibly he had been disturbed in the process of making his winter home.

Gulo lay low, or hunted very furtively, after that for some time, until it was little less dark in the east than it had been, and the gaunt tree-trunks were standing out a fraction from the general gloom. The moon had apparently nearly burnt itself out. Still, it yet appeared to be night.

Gulo was a long way out of his own hunting-district, and guessed that it was about time for him to get himself out of sight. He had a passionate hatred of the day, by the way, even beyond most night hunters.

On the way he smelt out and dug up a grouse beneath the snow.

Dawn found him, or, rather, failed to find him, hidden under a tangled mass that was part windfall, part brush-wood, and part snow. The place had belonged to a fox the night before, and that red worthy returned soon after dawn. He thrust an inquiring sharp muzzle inside, took one sniff, and, with every hair alift, retired in haste, without waiting to hear the villainous growl that followed him. The smell was enough for him--a most calamitous stink.

It snowed all that day, and things grew quieter and quieter, except in the tree-tops, where the wind spoke viciously between its teeth. When Gulo came out that evening, he had to dig part of the way, and he viewed a still and silent, white world, under a sky like the lid of a lead box, very low down. He stood higher against the tree-trunks than he had done the night before, and, though he did not know it, was safe from any horse, for the snow was quite deep. The cold was awful, but it did not seem to trouble him, as he slouched slowly southward.

There appeared to be nothing alive at all throughout this white land, but you must never trust to that in the wild. Things there are very rarely what they seem. For instance, Gulo came into a clearing, dim under the night sky, though it would never be dark that night. To the ear and the eye that clearing was as empty as a swept room. To Gulo's nose it was not, and he was just about to crouch and execute a stalk, when half the snow seemed to get up and run away. The runners were wood-hares. They had "frozen" stiff on the alarm from their sentries. But it was not Gulo who had caused them to depart. Him, behind a tree, they had not spotted. Something remained--something that moved. And Gulo saw it when it moved--not before. It was an ermine, a stoat in winter dress, white as driven snow. Then it caught sight of Gulo, or, more likely, the gleam of his eyes, and departed also.

Gulo slouched on, head down, back humped, tail low, a most dejected-looking, out-at-heels tramp of the wilderness.

Once he came upon a wild cat laying scientific siege to a party of grouse. The grouse were nowhere to be seen; nor was the wild cat, after Gulo announced his intention to break his neutrality. Gulo knew where the grouse were. He dug down into the snow, and came upon a tunnel. He dug farther, and came upon other tunnels, round and clean, in the snow. All the tunnels smelt of grouse, but devil a grouse could he find. He had come a bit early. It was as yet barely night, and he should have waited till later, when they would be more asleep. However, he dug on along the tunnels, driving the grouse before him. And then a strange thing happened. About three yards ahead of him the snow burst--burst, I say, like a six-inch shell, upwards. There was a terrific commotion, a wild, whirring, whirling smother, a cloud of white, and away went five birds, upon heavily beating wings, into the gathering gloom. Gulo went away, too, growling deep down inside of, and to, himself.

He was hungry, was Gulo. Indeed, there did not seem to be many times when he was not hungry. Also, being angry--not even a wild animal likes failure--he was seeking a sacrifice; but he had crossed the plain, which the night before had been as a nightmare desert to him, and the moon was up before his chance came.

He crossed the trail of the reindeer. He did not know anything about those reindeer, mark you, whether they were wild or semi-tame; and *I* do not know, though *he* may have done, how old the trail was. It was sufficient for him that they *were* reindeer, and that they had traveled in the general direction that he wanted to go. For the rest--he had the patience, perhaps more than the patience, of a cat, the determination of a bulldog, and the nose of a bloodhound. He trailed those reindeer the better part of that night, and most of the time it snowed, and part of the time it snowed hard.

By the time a pale, frozen dawn crept weakly over the forest tree-tops Gulo must have been well up on the trail of that herd, and he had certainly traveled an astonishing way. He had dug up one lemming--a sort of square-ended relation of the rat, with an abbreviated tail--and pounced upon one pigmy owl, scarce as large as a thrush, which he did not seem to relish much--perhaps owl is an acquired taste--before he turned a wild cat out of its lair--to the accompaniment of a whole young riot of spitting and swearing--and curled up for the day.

He was hungry when he went to sleep. Also, it was snowing then. When he woke up it was almost dark, and snowing worse than ever. If it could have been colder, it was.

While he cleaned himself Gulo took stock of the outside prospect, so far as the white curtain allowed to sight, and by scent a good deal that it did not. This without appearing outside the den, you understand. And if there had been any enemy in hiding, waiting for him outside, he would have discovered the fact then. He had many enemies, and no friends, had Gulo. All that he received from all whom he met was hate, but he gave back more than he got. In the lucid terms of the vernacular, he "was a hard un, if you like."

Nothing and nobody saw the wolverine leave that lair that was not his. He must have chosen one blinding squall of snow for the purpose, and was half a mile away, still on the track of the reindeer, before he showed himself--shuffling along as usual, a ragged, hard-bitten ruffian. And three hours later he came up with his prey.

Gulo knew it, but nobody else could have done. There were just the straight trees ahead, and all around the eternal white, frozen silence, and the snow falling softly over everything; but Gulo was as certain that there was the herd close ahead as he was that he was ravenous. And thereafter Gulo got to work, the peculiar work, a special devilish genius for which appears to be given to the wolverine.

He ceased to exist. At least, nothing of him was seen, not a tail, not an eye-gleam. Yet during the next two hours he learnt everything, private and public, there was to be learnt. Also, he had been over the surroundings almost to a yard. Nothing could have escaped him. No detail of risk and danger, of the chance of being seen even, had been overlooked; for he was a master at his craft, the greatest master in the wild, perhaps. The wolf? My dear sirs, the wolf was an innocent suckling cub beside Gulo, look you, and his brain and his cunning were not the brain and the cunning of a beast at all, but of a devil.

When, after a very long time, he reappeared upon his original track, it was as a dark blotch, indistinguishable from a dozen other dark blots of moon-shadow, creeping forward belly-flat in the snow. This belly-creep, hugging always every available inch of cover, he kept up till he came to a big clearing, and--there were the reindeer. At least, there was one reindeer, a doe, standing with her back towards him--a quite young doe. The rest were half-hidden in the snow, which they had trampled into a maze of paths in and out about the clearing, which was, in fact, what is called their "yard."

A minute of tense silence followed after Gulo had got as close as he could without being seen. Then he rushed.

The reindeer swung half-round, gave one snort, and a great bound. But Gulo had covered half the intervening space before she knew, and when she bounded it was with him hanging on to her.

Followed instantly a wild upspringing of snorting beasts, and a mad, senseless stampede of floundering deer all round and about the clearing--a fearful mix-up, somewhere in the midst of which, half-hidden by flying, finely powdered snow, Gulo did his prey horribly to death.

There was something ghastly about this murder, for the deer was so big, and Gulo comparatively small. The fearful work of his jaws and his immense strength seemed wrong somehow, and out of all proportion to his size. This remarkable power of his jaws had that sinister disproportion only paralleled by the power of the jaws of a hyena; indeed, his teeth very much resembled a hyena's teeth.

With the deer rushing all around him, Gulo fed, ravenously and horribly, but not for long. A new light smoldered in his eyes now as he lifted his carmine snout, and one saw that, for the moment, the beast was mad, crazed with the lust of killing, seeing red, and blinded by blood.

Then the massacre began. It was not a hunt, because each deer, thinking only of itself, feared to break from the trodden mazy path of the "yard," and risk the slow, helpless, plunging progress necessary in the deep snow. Wherefore panic took them all over again, and they dashed, often colliding, generally hindering each other, hither and thither, up and down the paths of the "yard" with the hopeless, helpless, senseless, blind abandon of sheep. The result was a shambles.

This part we skip. Probably--nay, certainly--Nature knows best, and is quite well aware what she is up to, and it is perhaps not meant that we should put her in the limelight in her grisly moods. Suffice it to say that

Gulo seemed to stop at length, simply because even he could not "see red" forever, and with exhaustion returned sense, and with sense--in his case--in-born caution. He removed, leaving a certain number of reindeer bleeding upon the ground. Some of them were dead.

In an hour dawn would be conspiring to show him up before the world, and he was not a beast sweet to look upon at that moment--indeed, at any moment, but less so now.

Now, it is surprising how far a wolverine can shift his clumsy-looking body over snow in an hour, especially if he has reasons. This one had good reasons, and he was no fool. He knew quite well the kind of little hell he had made for himself behind there, and he did not stay to let the snow cover him. He traveled as if he were a machine and knew no fatigue; and the end of that journey was a hole in a hollow among rocks.

Dawn was throwing a wan light upon all things when he thrust his short, sharp muzzle inside that hole, to be met by a positively hair-raising volley of rasping, vicious growls.

He promptly ripped out a string of ferocious, dry, harsh growls in return, and for half-a-minute the air became full of growls, horrible and blood-curdling, each answering each.

Then he lurched in over the threshold, and coolly dodged a thick paw, with tearing white claws, that whipped at him with a round-arm stroke out of the pitch-darkness. The stroke was repeated, scraping, but in nowise hurting his matted coat, as he rose on his hindlegs and threw himself upon the striker.

Followed a hectic thirty seconds of simply diabolical noises, while the two rolled upon the ground, grappling fiendishly in the darkness. Then they parted, got up, growled one final roll of fury at each other, fang to fang, and, curling up, went to sleep. But it was nothing, only the quite usual greeting between Gulo and his wife. They were a sweet couple.

There appeared to be no movement, or any least sign of awakening, on the part of either of the couple between that moment and sometime in the afternoon, when, so far as one could see, Gulo suddenly rolled straight from deep sleep out on to the snow, and away without a sound, at his indescribable shamble and at top speed.

Mrs. Gulo executed precisely the same amazing maneuver, and at exactly the same moment, as far as could be seen, on the other side, and shuffled off into the forest. They gave no explanation for so doing. They said never a word--nothing. One moment they were curled up, asleep; the next they had gone, scampered, apparently into the land of the spirits, and were no more. Nor did there seem to be any reason for this extraordinary conduct except--except---- Well, it is true that a willow-grouse, white as the snowy branch he sat upon, *did* start clucking somewhere in the dim tree regiments, a snipe did come whistling sadly over the tree-tops, and a raven, jet against the white, did flap up, barking sharply, above the pointed pine-tops; but that was nothing--to us. To the wolverines it was everything, a whole wireless message in the universal code of the wild, and they had read it *in their sleep*. Through their slumbers it had spelt into their brains, and instantly snapped into action that wonderful, faultless machinery that moved them to speed as if automatically.

Then the chase began, grim, steady, relentless, dogged--the chase of death, the battle of endurance.

A pause followed after the vanishing of the hated wolverines. A crow lifted on rounded vans, marking their departure, and it was seen. A blackcock launched from a high tree with a whir and a bluster like an aeroplane, showing their course, and it was noted. An eagle climbed heavily and ponderously over the low curtain of the snow mist, pointing their way, and it was followed. All the wild, all the world, seemed to be against the wolverines. The brigands were afoot by day. The scouts were marking their trail.

Then a lynx, moving with great bounds on his huge swathed paws, shot past between the iron-hard tree-boles; a fox followed, scudding like the wind on the frozen crest; a hare, white as a waste wraith, flashed by, swift as a racing white cloud-shadow; a goshawk screamed, and drew a straight streaking line across a glade. And then came the men, side by side, deadly dumb, with set faces, the pale sun glinting coldly cruel upon the snaky, lean barrels of their slung rifles, moving with steady, fleet, giant strides on their immense spidery ski that were eleven feet long, which whispered ghostily among the silent aisles of Nature's cathedral of a thousand columns. The Brothers were on the death-trail of Gulo at last; the terrible, dreaded Brothers, who could overtake a full-grown wolf in under thirty minutes on ski, and whose single bullet spelt certain death. Now for it; now for the fight. Now for the great test of the "star" wild outlaw against the "star" human hunters--at last. The reindeer were to be avenged.

Then Time took the bit of silence between his teeth and seconds became hours, and minutes generations.

No sound made the wolverines as they rolled along in Indian file, except for the soft whisper of the snow underfoot.

No noise encompassed the Brothers as they sped swiftly side by side over the glittering white carpet, save for the slither of the snow under their weight.

All the wild seemed to be standing still, holding its breath, looking on, spell-bound; and save for the occasional crash of a collapsing snow-laden branch, sounding magnified as in a cave, all the forest about there was as still as death.

Half-an-hour passed, and Gulo flung his head around, glancing over his shoulder a little uneasily, but with never a trace of fear in his bloodshot eyes. Then he grunted, and the two fell apart silently and instantly, gradually getting farther and farther from each other on a diverging course, till his wife faded out among the trees. But never for an instant did either of them check that tireless, deceptive, clumsy, rolling slouch, that slid the trees behind, as telegraph-poles slide behind the express carriage window.

Half-an-hour passed, and one of the Brothers, peering up and along the trail a little anxiously, saw the forking of the line ahead. Then he grunted, and the two promptly separated without a word, gradually increasing the distance between them on the widening fork till they were lost to each other among the marshaled trunks. But never for an instant did they relax that swift, ghostly glide on the wonderful ski, that slid the snow underfoot as a racing motor spins over the ruts.

An hour passed. Sweat was breaking out in beads upon the faces of the Brothers, now miles apart, but both going in the same general direction over the endless wastes of snow, and upon their faces was beginning to creep the look of that pain that strong men unbeaten feel who see a beating in sight; but never for a moment did they slacken their swift, mysterious glide.

An hour passed. Foam began to fleck the evilly uplifted lips glistening back to the glistening fangs of the wolverines, now miles apart, but still heading in the same general line, and upon their faces began to set a look of fiends under torture; but never for a moment did they check their indescribable shuffling slouch.

After that all was a nightmare, blurred and horrible, in which endless processions of trees passed dimly, interspersed with aching blanks of dazzling white that blinded the starting eyes, and man and beast stumbled more than once as they sobbed along, forcing each leg forward by sheer will alone.

At last, on the summit of a hog-backed, bristling ridge, Gulo stopped and looked back, scowling and peering under his low brows. Beneath him, far away, the valley lay like a white tablecloth, all dotted with green pawns, and the pawns were trees. But he was not looking for them. His keen eyes were searching for movement, and he saw it after a bit, a dot that crept, and crept, and crept, and--*stopped*!

Gulo sat up, shading his eyes against the watery sun with his forepaws, watching as perhaps he had never watched in his life before.

For a long, long while, it seemed to him, that dot remained there motionless, far, far away down in the valley, and then at length, slowly, so slowly that at first the movement was not perceptible, it turned about and began to creep away--creep, creep, creep away by the trail it had come.

Gulo watched it till it was out of sight, fading round a bend of the hills into a dark, dotted blur that was woods. Then he dropped on all fours, and breathed one great, big, long, deep breath. That dot was the one of the Brothers that had been hunting him.

And almost at the same moment, five miles away, his wife had just succeeded in swimming a swift and ice-choked river. She was standing on the bank, watching another dot emerge into the lone landscape, and that dot was the other one of the Brothers.

They had failed to avenge the reindeer, and the wolverines were safe. Safe? Bah! Wild creatures are never safe. Nature knows better than that, since by safety comes degeneration.

There was a warning--an instant's rustling hissing in the air above--less than an instant's. But that was all, and for the first time in his life--perhaps because he was tired, fagged--Gulo failed to take it. And you must never fail to take a warning if you are a wild creature, you know! There are no excuses in Nature.

Retribution was swift. Gulo yelled aloud--and he was a dumb beast, too, as a rule, but I guess the pain was excruciating--as a hooked stiletto, it appeared, stabbed through fur, through skin, deep down through flesh, right into his back, clutching, gripping vise-like. Another stiletto, hooked, too, worse than the first one, beat at his skull, tore at his scalp, madly tried to rip out his eyes. Vast overshadowing pinions--as if they were the wings of Azrael--hammered in his face, smothering him, beating him down.

Ah, but I have seen some fights, yet never such a fight as that; and never again do I want to see such a fight as the one between Gulo and the golden eagle that made a mistake in his pride of power.

All the awful, cruel, diabolical, clever, devilish, and yet almost human fury that was in that old brute of a Gulo flamed out in him at that moment, and he fought as they fight who go down to hell. It was frightful. It was terrifying. Heaven alone knows what the eagle thought he had got his claws into. It was like taking hold of a flash of forked lightning by the point. It was--great!

Still, flight *is* flight, and lifting-power is lifting-power. Gulo, even Gulo, could not get over that. He could not stop those vast vans from flapping; and as they flapped they rose, the eagle rose, he--though it was like the skinning of his back alive--rose too, wriggling ignominiously, raging, foaming, snapping, kicking, but--he rose.

Slowly, very slowly, the great bird lifted his terrible prey up and up--ten, twenty, thirty, forty feet, but no higher. That was the limit of his lift, the utmost of his strength; and at that height parallel with the ridge, he began to carry the wolverine along, the wolverine that was going mad with rage in his grasp.

It was a mistake, of course--a mistake for the wolverine to be out on the open ridge in stark daylight; another mistake for the eagle, presuming on his fine, lustful pride of strength, to attack him.

And then suddenly Gulo got his chance. It hit him bang in the face, nearly blinding him as it passed--the tree-top. Like lightning Gulo's jaws clashed shut upon it, his claws gripped, and--he thought his back was going to come off whole. But he stuck it. He was not called Gulo the Indomitable for nothing. And the eagle stopped too. He had to, for he would not let go; nor would Gulo.

An awful struggle followed, in the middle of which the pine-top broke, gave way, and, before either seemed to know quite what was happening, down they both came, crashing from branch to branch, to earth.

The fall broke the king of the birds' hold, but not the fighting fury of the most hated of all the beasts. He rose up, half-blind, almost senseless, but mad with rage beyond any conception of fury, did old Gulo, and he hurled himself upon that eagle.

What happened then no man can say. There was just one furious mix-up of whirling, powdered snow, that hung in the air like a mist, out of which a great pinion, a clawing paw, a snapping beak, a flash of fangs, a skinny leg and clutching, talons, a circling bushy tail appeared and vanished in flashes, to the accompaniment of stupendous flappings and abominably wicked growls.

That night the lone wolf, scouting along the ridge-top, stopped to sniff intelligently at the scattered, torn eagle's feathers lying about in the trampled snow, at the blood, at the one skinny, mailed, mightily taloned claw still clutching brown-black, rusty fur and red skin; at the unmistakable flat-footed trail of Gulo, the wolverine, leading away to the frowning, threatening blackness of the woods. He could understand it all, that wolf. Indeed, it was written there quite plainly for such as could read. He read, and he passed on. He did not follow Gulo's bloody trail. No--oh, dear, no! Probably, quite probably, he had met Gulo the Indomitable before, and--was not that enough?

II

BLACKIE AND CO.

Blackie flung himself into the fight like a fiery fiend cut from coal. He did not know what the riot was about--and cared less. He only knew that the neutrality of his kingdom was broken. Some one was fighting over his borders; and when fighting once begins, you never know where it may end! (This is an axiom.) Therefore he set himself to stop it at once, lest worse should befall.

He found two thrushes apparently in the worst stage of d.t.'s. One was on his back; the other was on the other's chest. Both were in a laurel-bush, half-way up, and apparently they kept there, and did not fall, through a special dispensation of Providence. Both fought like ten devils, *and both sang*. That was the stupefying part, the song. It was choked, one owns; it was inarticulate, half-strangled with rage, but still it *was* song.

A cock-chaffinch and a hen-chaffinch were perched on two twigs higher up, and were peering down at the grappling maniacs. Also two blue titmice had just arrived to see what was up, and a sparrow and one great tit were hurrying to the spot--all on Blackie's "beat," on Blackie's very own hunting-ground. Apparently a trouble of that kind concerned everybody, or everybody thought it did.

Blackie arrived upon the back of the upper and, presumably, winning thrush with a bang that removed that worthy to the ground quite quickly, and in a heap. The second thrush fetched up on a lower branch, and by the time the first had ceased to see stars he had apparently regained his sanity. He beheld Blackie above him, and fled. Perhaps he had met Blackie, professionally, before, I don't know. He fled, anyway, and Blackie helped him to flee faster than he bargained for.

By the time Blackie had got back, the first thrush was sitting on a branch in a dazed and silly condition, like a fowl that has been waked up in the night. Blackie presented him with a dig gratis from his orange dagger, and he nearly fell in fluttering to another branch. And Blackie flew away, chuckling. He knew that, so far as that thrush was concerned, there would be no desire to see any more fighting for some time.

But, all the same, Blackie was not pleased. He was worked off his feet providing rations for three ugly youngsters in a magnificently designed and exquisitely worked and interwoven edifice, interlined with rigid cement of mud, which we, in an off-hand manner, simply dismiss as "A nest." The young were his children; they might have been white-feathered angels with golden wings, by the value he put on them. The thrush episode was only a portent, and not the first. He had no trouble with the other feathered people he tolerated on his beat.

Blackie went straight to the lawn. (Jet and orange against deep green was the picture.)

Now, if you and I had searched that dry lawn with magnifying-glasses, in the heat of the sun, there and then, we should not have found a single worm, not the hint or the ghost of one; yet that bird took three long, low hops, made some quick motion with his beak--I swear it never seemed to touch the ground, even, let alone dig---executed a kind of jump in the air--some say he used his legs in the air--and there he was with a great, big, writhing horror of a worm as big as a snake (some snakes).

Thrushes bang their worms about to make them see sense and give in; they do it many times. Blackie banged his giant only a little once or twice, and then not savagely, like a thrush. Also, again, he may or may not have used his feet. Moreover, he gave up two intervals to surveying the world against any likely or unlikely stalking death. Yet that worm shut up meekly in most unworm-like fashion, and Blackie cut it up into pieces. The whole operation took nicely under sixty seconds.

Blackie gave no immediate explanation why he had reduced his worm to sections. It did not seem usual. Instead, he eyed the hedge, eyed the sky, eyed the surroundings. Nothing seemed immediately threatening, and he hopped straight away about three yards, where instantly, he conjured another and a smaller worm out of nowhere. With this unfortunate horror he hopped back to the unnice scene of the first worm's decease, and carved that second worm up in like manner. Then he peeked up all the sections of both worms, packing them into his beak somehow, and flew off. And the robin who was watching him didn't even trouble to fly down to the spot and see if he had left a joint behind. He knew his blackbird, it seemed.

Blackie flew away to his nest, but not to a nest in a hedge. To dwell in a hedge was a rule of his clan, but the devil a rule did he obey. Nests in hedges for other blackbirds, perhaps. He, or his wife, had different notions. Wherefore flew he away out into the grass field behind the garden. Men had been making excavations there, for what mad man-purpose troubled him not--digging a drain or something. No matter.

Into the excavation he slipped---very, very secretly, so that nobody could have seen him go there--and down to the far end, where, twelve feet below the surface, on a ledge of wood, where the sides were shored with timber, his mate had her nest. Here he delivered over his carved joints to the three ugly creatures which he knew as his children and thought the world of, and appeared next flying low and quickly back to the garden. That is to say, he had contrived to slip from the nest so secretly that that was the first time he showed.

A sparrow-hawk, worried with a family of her own, took occasion to chase him as he flew, and he arrived in among the young lime-trees that backed the garden, switchbacking--that was one of his tricks of escape, made possible by a long tail--and yelling fit to raise the world. The sparrow-hawk's skinny yellow claw, thrust forward, was clutching thin air an inch behind his central tail-feathers, but that was all she got of him--just thin air. There was no crash as he hurled into the green maze; but she, failing to swerve exactly in time, made a mighty crash, and retired somewhat dazed, thankful that she retained two whole wings to fly with. There is no room for big-winged sparrow-hawks in close cover, anyway, and Blackie, who was born to the leafy green ways, knew that.

Blackie's yells had called up, as if by magic, a motley crowd of chaffinches, hedge-sparrows, wrens, robins, &c., from nowhere at all, and they could be seen whirling in skirmishing order--not too close--about the retreating foe. Blackie himself needed no more sparrow-hawk for a bit, and preferred to sit and look on. If the little fools chose to risk their lives in the excitement of mobbing, let them. His business was too urgent.

Twenty lightning glances around seemed to show that no death was on the lurk near by. Also, a quick inspection of other birds' actions--he trusted to them a good deal--appeared to confirm this.

Then he flew down to the lawn, and almost immediately had a worm by the tail. Worms object to being so treated, and this one protested vigorously. Also, when pulled, they may come in halves. So Blackie did not pull *too* much. He jumped up, and, while he was in the air, scraped the worm up with his left foot, or it may have been both feet. The whole thing was done in the snap of a finger, however, almost too quickly to be seen.

The worm, once up, was a dead one. Blackie seemed to kill it so quickly as almost to hide the method used. In a few seconds more it was a carved worm in three or four pieces--an unnice sight, but far more amenable to reason that way.

Blackie was in rather long grass, and nerve-rackingly helpless, by the same token. He could not see anything that was coming. Wherefore every few seconds he had to stand erect and peer over the grass-tops. It made no difference to the worm, however; it was carved just the same.

Blackie now hopped farther on in search of more worms, but found a big piece of bread instead. It was really the hen-chaffinch who found the bread, and he who commandeered it from her. Now he disclosed one fact, and that was that bread would do for his children as well as worms. Anyhow, he stuffed his beak about half up with bread, returned for the pieces of worm, collected these, and retired up into the black cover of a fir-tree.

No doubt he was expected to go right home with that load, then and there. Even a cock-blackbird with young, however, must feed, and, if one judged by the excess amount of energy--if that were possible--used up, must feed more than usual. That seemed to be why he hid his whole load in the crook of a big bough, and, returning to the lawn, ate bread--he could wait to catch no worms for his own use, it appeared--as fast as he could. Three false alarms sent him precipitantly into his tree upon this occasion, and one real alarm--a passing boy--caused a fourth retreat.

These operations were not performed in a moment, and by the time he got back to his nest--mind, he had to contrive to approach it so that he was seen by nobody, and his was a conspicuous livery, too--his children appeared to be in the last stages of exhaustion. That, however, is young birds all over; they expect their parents to be mere feeding-machines, guaranteed to produce so many meals to the hour, and hang the difficulties and the risks.

There was no sign of Blackie's wife. Presumably she was working just as hard on her own "beat" as he had been on his--their hunting-grounds were separate, though they joined--and would soon be back.

Blackie did not wait. He managed again the miracle of getting away from his nest without appearing to do so, and next turned up on the summerhouse roof. Fatherlike, he thought he had done enough for a bit, and would enjoy a "sunning reaction" on the summerhouse roof. It was rather a good place, a look-out tower from which he could slip over the side into the hedges, which met at the corner where it was, if trouble turned up.

Trouble did turn up, but not quite what he had expected. He had been sitting there, wing-stretching, leg-stretching, and "preening" his feathers, and had finally left off just to sit and do nothing, when--lo! his wife popped up over the side without warning, and right upon him.

She was very dark brown, not black, and had a paler throat than the palish throat of most hen blackbirds--nearly white, in fact. She said nothing. Nor did Blackie, but he looked very uncomfortable. She did more than say nothing. She went for him, beak first, and very angrily indeed; and he, not waiting to receive her, fled down to the lawn, and began worm-hunting for dear life. A whole lecture could not have said more.

Mrs. Blackie remained on the roof for about a minute, looking round, and then flew off to her own hunting-ground. She was wilder and less trusting of the world than Blackie, and did not care for his lawn in full view of the house windows. And Blackie did not even stop his work to watch her go. Apparently they had, previously in their married life, arrived at a perfect understanding.

This time, too, Blackie got a big and a small worm. The small he coiled like a rope, and held up towards the base of his beak; the big he carved up into sections, which he held more towards the tip. The large ones, it seemed, were too awkward and lively simply to carry off rolled up whole.

The journey that followed was a fearful one in Blackie's life, for he met half-way the very last foe in the world he was expecting--namely, an owl. Truly, it was a very small owl, scarce bigger than himself; but it was an owl, and, like all its tribe, armed to the teeth. Men called it a little owl. That was its name--little owl. Blackie didn't care what men called it; he knew it only as one of the hundred or so shapes that death assumed for his benefit.

Just at that time it happened to be cloudy, and little owls often hunt by day. But how was Blackie to know that, little owls being a comparatively new introduction into those parts?

Blackie screamed and fled. The owl did not scream, but fled, too--after Blackie. Blackie had no means of judging how close *that* foe was behind by the whir of its wings. Owls' wings don't talk, as a rule; they have a patent silencer, so to speak, in the fluffy-edged feathers. Therefore Blackie was forced to do his best in breaking the speed record, and trust to luck.

It was a breathless and an awful few seconds, and it seemed to him like a few hours. The owl came up behind, going like a cloud-shadow, and about as fast, and Blackie, glancing over his shoulder, I suppose, yelled afresh. The terror was so very close.

Then Blackie remembered another excavation, just like the one his nest was in, a little off his course to the left, and he tacked towards it, twisting his course wonderfully, thanks to the long tail. And the owl lost a foot on the turn. I think it was expecting Blackie to make for the hedge at all costs. But, be that as it may, that foot was never made up again, for Blackie vanished into the trench next instant, like a blown-out light, and, though the hunter searched for him carefully, he never put in an appearance again while that owl was within sight of the place.

"The owl lost a foot on the turn"—*Page 34*

[Illustration: "The owl lost a foot on the turn"]

All signs of uproar on the passing of the little owl had died down some time before Blackie turned up again, and then it was in the garden, so he must have got from the tunnel unseen.

He still hung gamely to the food for his young, and now made another attempt to deliver that food where it belonged. He was half-way there, indeed, before he saw the boys--three boys--with two rows of birds' eggs threaded on strings. They were passing so close to the trench that one nearly fell into it, and, of course, any one could see that they were bird-nesting.

Blackie swerved off sharply to the far hedge, his heart nearly bursting with anxiety, little knowing that the boys had never even thought of looking in the trench for nests. It seemed the last place in the world to find one. It may have been, moreover, that he feared that his wife was home, in which case she might have lost her head, and, dashing out with a scream, "blown the whole gaff," as they say in the vernacular.

Apparently madam was not at home, by good luck, for the boys passed, and Blackie once more executed the magic of getting into his nest without seeming to do so. And here he stayed. Dusk was setting in, and his young were fourteen days old. They showed it in their disobedience, and were not in the least inclined to keep as quiet as they should, considering their father had but just warned them, in his own way, to "lie doggo," because of the gray shape he had seen sliding out into the field, a gray shape which was a cat.

They were more like thrushes than blackbirds, those youngsters, with their speckly fawn breasts; and they were not like the adults of either in their frog-like attitudes and heavy ways. Frankly, they were not beautiful, even at that stage; and a fortnight before, when they had been larger than the eggs they had come out of, they were positively reptilian and repulsive.

Blackie said the blackbird equivalent to "Be quiet, you little fools!" as quietly and as sternly as he could three or four times, and perched on the top of a wheelbarrow to watch the gray shadow which was a cat. That sudden death, however, was more afraid of the open than Blackie, even, and, moreover, wasn't expecting blackbirds' nests in the middle of fields. It turned back; and at length Mrs. Blackie, who had been on a general survey round about to see what foes of the night were on the move--and a fine hubbub she had made in the process--came home, reporting all well.

Then they slept; at least, they ceased to be any further seen or heard. Think, however, how you would sleep if every few minutes you could hear sounds in your house one-third of which were probably the noises of a burglar. Think, also, how you would feel if you knew that that burglar was a murderer, and that that murderer was, in all likelihood, looking for you, or some one just like you. Yet those birds were happy enough, I fancy.

It was barely pale gray. It was cold and wan and washed, and wonderfully clean and sweet, and wet with dews, when a lark climbed invisibly into the sky and suddenly burst into song, next morning. There was something strange and out of place, in a way, in this song, breaking out of the night; and as it and another continued to break the utter silence for ten minutes, it seemed rather as if it were still night, and not really dawn at all. Dawn appeared to be waiting for something else to give it authority, so to speak, and at the end of ten minutes that something else came--the slim form of Blackie, streaking, phantom-like, through the mist from the trench out in the field to the summerhouse in the garden. Here, mounted upon the very top, he stood for a moment, as one clearing his throat before blowing a bugle, and then, full, rich, deep, and flute-like, he lazily gave out the first bars of his song. Instantly, almost as if it had been a signal, a great tit-mouse sang out, "Tzur ping-ping! tzur ping-ping!" in metallic, ringing notes; a thrush struck in with his brassy, clarion challenge, thrush after thrush taking it up, till, with the clear warble of robin and higher, squeaking notes of hedge-sparrow and wren joining in, the wonderful first bars of the Dawn Hymn of the birds rolled away over the fields to the faraway woods, and beyond.

Blackie sang on for a bit, in spite of the fact that people said that it was not considered "the thing" for a blackbird with such domestic responsibilities to sing. And two other blackbirds helped him to break the man-made rule.

As a matter of fact, I fancy he was not taking chances upon the ground while the mist hung to cover late night prowlers, for as soon as the gay and gaudy chaffinches had stuck themselves up in the limes and the sycamores, and started their own smashing idea of song, he was down upon the lawn giving the early worm a bad time.

Then it was that he heard a rumpus that shot him erect, and sent his extraordinarily conspicuous orange dagger of a beak darting from side to side in that jerky way of listening that many birds affect.

"Twet-twet-et-et-et! twet! twet-twet-twet-et-et-twet!" came the unmistakable voice of one in a temper, scolding loudly. And he knew that scold--had heard it before, by Jove! And who should know it if not he, since it was the voice of his wife?

Perhaps he heaved a sigh as he rose from the deliciously cool, wet lawn--where it was necessary to take long, high hops if you wanted to avoid getting drenched--and winged his way towards the riot. His wife was calling him, and it came from the other side of the garden, *her* side, behind the house. Perhaps it was a cat, or a rat, or something. Anything, almost, would set her on like that if experience, plus the experience of blackbirds for hundreds of generations working blindly in her brain--and not the experience of books--had taught her that the precise creature whom she saw was a danger and a menace to young blackbirds.

All the same, when Blackie arrived he *was* surprised, for all that he saw was a grayish bird with "two lovely black eyes," not by any means as large as a blackbird. When it flew it kept low, with a weak and peculiar flight that was deceiving; and when Mrs. Blackie, following it, and yelling like several shrews, got too close, it turned its head, and said, "Wark! wark!" in a harsh and warning way.

Blackie joined in with his deeper "Twoit-twoit-twoit!" just by way of lending official dignity to the proceedings. Whereupon his wife, feeling that he had backed her up, redoubled her excitement and shrill abuse.

And they spent two solid hours at this fool's game, helped by a robin, a blue tit, and a chaffinch or two--the chaffinch must have his finger in every pie--following that gray bird from nowhere, while it moved about the garden in its shuffling flight, or alternately sat and scowled at them. But it must be admitted that Blackie himself looked rather bored, and might have gone off for breakfast any time, if he had dared.

As a matter of fact, however, the bird did not stand upon the Register of Bad Deeds as being a terror of even the mildest kind of blackbirds. Red-backed shrike was her name, female was her sex, and from Africa had she come. Goodness knows where she was going, but not far, probably; and the largest thing in the bird line she appeared able to tackle was something of the chaffinch size. But, all the same, Mrs. Blackie seemed jolly well certain that she knew better.

Then arrived the bombshell.

One of the Blackie youngsters, stump-tailed, frog-mouthed, blundering, foolish, gawky, and squawking, landed, all of a heap, right into the very middle of the picnic-party.

Mrs. Blackie very nearly had a fit on the spot, and the shrike judged that the time had about arrived for her to quit that vicinity.

Blackie himself, to do him justice, kept cool enough to do nothing. Wives will say that he was just husband all over, but there were reasons abroad. One of them shot past Blackie, who was low down, a second later and a yard away, and had he not been absolutely still, and therefore as invisible as one of the most conspicuous of birds in the wild can be, he would have known in that instant, or the next, what lies upon the other side of death.

Another reason shot through the lower hedge, and, both together, they fell upon the young bird.

They were the cat of the house and her half-grown kitten, and they were upon the unhappy youngster before you could shout, "Murder!"

What followed was painful. Mrs. Blackie went clean demented. Blackie went--not so demented. (It always appeared to me that his was a more practical mind than his wife's, perhaps because he wore a more conspicuous livery.) Mrs. Blackie kept passing and repassing the cats' backs, flying from bough to bough, and sometimes touching and sometimes not touching them. It was useless, of course, this pathetic charging; and it was foolish.

Blackie charged, too, but not within feet.

Suddenly the old cat, who had had one eye upon Mrs. Blackie the whole time, sprang up and struck quickly twice. There was a chain of shrieks from Mrs. Blackie, and down she went in the grip of the clawed death. She never got up again.

What had happened was simple enough. One of the laborers working on the trench, knowing of the nest, had, out of curiosity, approached a little too close, when the bevy of youngsters, being ready to fly, but not knowing it before this great fright, burst apart at his approach like a silent cannon cracker. The fear showed them they could use their wings.

All three had made, flying low and weakly, for the nearest hedge, which was the garden hedge--two to the side which comprised their father's hunting-ground, and one to the side dominated by their ma. And the old cat and her kitten had seen them coming, and had given chase.

Blackie discovered his remaining progeny sitting about in bushes, squawking, a few minutes later, when he returned, somewhat agitated to his precious lawn, and there he promptly proceeded to feed them. The task was such a large one, and took so long, and so many worms had to be cut up, and so much bread, and, I may say, when all else failed, so many daisies had to be picked, before he finally silenced their ceaseless, craving remarks, that, by the time he had finished feeding himself and had a clean up, something of the pain of the tragedy had gone from him.

And fine fat blackbirds he made of those youngsters, too, in the end, I want to tell you, for he stuck to 'em like a brick.

III

UNDER THE YELLOW FLAG

A little past noon each day the sun covered a crack between two boards on the summerhouse floor, and up through that aperture, for three days, had come a leggy, racy-looking, wolfish black spider. Each day, as it grew hotter, she extended her sphere of jerky investigation, vanishing down the crack again when the sun passed from it.

To-day she prolonged her roamings right up the wall of the summerhouse and along a joist bare of all save dust, and--well, the spider walked straight on, moving with little jerks as if by intermittent clockwork, and she seemed to stroll right on top of the wasp lying curled up on her side. Only when one of the latter's delicate feelers shifted round towards her, as though in some uncanny way conscious of her approach, did she leap back as if she had touched an electric wire. Then she froze--flat. The wasp was lying curled up, as we have said, upon her side, her head tucked in, her wings drawn down, her jaws tight shut upon a splinter of wood. She had been there half-a-year, asleep, hibernating, and in that state, without any other protection than the summerhouse roof and walls, had survived the frosts of winter.

The wasp did not move further.

The spider appeared to be taking things in, measuring her chances, weighing the risk against her famished hunger--possibly her late husband had been her last meal, months ago--marking the vital spot upon her prey, aiming for the shot, which must be true, for one does not miss in attacking a wasp--and live. Only, she would not have risked it at all, perhaps, if the wasp had seemed alive, or more alive, at any rate.

Then came the shot--one cannot in justice call it a spring; it was too instant to be termed that. The spider simply was upon the wasp without seeming to go there; but the wasp was not there, or, rather, her vital spot wasn't. She had kicked herself round on her side, like a cart-wheel, lying flat, with her feet, and the spider's jaws struck only hard cuirass. Before the spider, leaping back, wolf-like, could lunge in her lightning second stroke, the wasp was on her feet, a live thing, after all.

The warmth had been already soaking the message of spring into her cold-drugged brain, and now this sudden attack had finished what the warmth had begun. She was awake, on her feet, a live and dangerous proposition; groggy, it is true; dazed, half-working, so to speak; but a force to be reckoned with--after half-a-year. And one saw, too, at a glance that she was different from ordinary wasps--would make two, in fact, of any ordinary wasp; and her great jaws looked as if they could eat one and comfortably deal with more; whilst her dagger-sting, now unsheathed and ready--probably for the first time--could deliver a wound twice as deep and deadly as the ordinary wasp. She was, in short, a queen-wasp; a queen of the future, if Fate willed; a queen as yet without a kingdom, a sovereign uncrowned, but of regal proportions and queenly aspect, for all that; for in the insect world royalties are fashioned upon a super-standard that marks them off from the common herd.

The spider hesitated. She knew the danger of the stripes of yellow--the yellow flag, so to speak. The fear of it is upon every insect that lives. At the same time, the queen was undoubtedly yet numb.

Antagonists decide in the insect world like a flash of light, and quick as thought they act. The spider attacked now so quickly that she seemed to have vanished, and she met--jaws. Back she shot, circled, shot in again, and she met--sting!

It was never clear whether that sting went home. The spider did. She fell--fell plump to the floor, only not breaking what spiders have in place of a neck because of the fact that, being a spider, she never moved anywhere, not even upon a spring, without anchoring a line of web down first. Therefore, an inch from the ground, she fetched up with a jerk upon the line that she had anchored up on the joist, spun round, let herself drop the rest of the way, and ran into the crack between the boards of the floor. Goodness knows if she lived.

The wasp, with that extremely droll, lugubrious look on her long, mask-like face which makes the faces of insects so funny and uncanny, like pantomime masks, sat down as if nothing had happened, apparently to scheme out the best way to possess herself of a kingdom and become a queen in fact as well as in name. Really, she was cleaning herself--combing her antennae with her forelegs, provided with bristle hairs for the purpose, scraping and polishing her wings, as if they did not already shine like mother-o'-pearl, and washing her quaint face.

She was still rather groggy from the effects of her long sleep and the cold endured--it is a wonder how she had stood the latter at all--and when, with a subdued inward sort of hum, she finally launched herself in flight, she nearly fell to the ground before righting herself and flying in a zigzag heavily across the lawn.

A cock-chaffinch up in the limes saw her, and condescending at last to break his song, described a flashing streak of wine-red breast and white wing-bars in the sun. He appeared to recognize her sinister yellow shield in time, however, and returned to his perch with a flourish, leaving the wasp to go on and begin dancing up the wall of the house till she came to the open window. Here she vanished within.

The sunlight sat on the floor of the room inside, and the baby sat in the sunlight; and the wasp, apparently still half-awake, went, or, rather, nearly tumbled, and sat beside the baby.

They made an odd picture there--the golden sun, the sunny, golden-headed baby, and that silent, yellow she-devil, crawling, crawling, crawling, with her narrow wings gleaming like gems.

Then the child put out her chubby hand to seize that bright-yellow object--how was she to know that it was the yellow signal of danger in the insect world that she saw? And, of course, being a baby, she was going to stuff it into her mouth. But Fate had use for that wasp--perhaps for that baby. Wherefore there was a little scream, a pair of woman's arms swept down and whisked that baby into the air, and a high-heeled shoe whisked the astonished wasp into a corner. Here she swore savagely, vibrating her head with tremendous speed in the process, rose heavily and menacingly, made to fly out, hit the upper window, which was shut, and which she could not see, but felt, and fell to the floor again, where she apparently had brain-fever, buzzing round and round on her back like a top the while.

And then, rising suddenly, the queen flew away, hitting nothing in the process, but getting through the lower and open part of the window. She seemed anxious to make sure of not getting into the house again. She flew right away, rising high to top the garden hedge, and dropping low on the far side, to buzz and poke about in and out, up along the hedge-bank that bordered the hayfield.

She flew as one looking for something, and every insect in her way took jolly good care--in the shape of scintillating streaks and dashes--to get out of it. The mere sight of that yellow-banded cuirass shining in the sun was apparently quite enough for them--most of them, anyway. As a matter of fact, she was looking for a site for a city. She had ambition, and would found her a city, a city of her very own, with generous streets at right angles, on the American plan; and she would be queen of it. It was a big idea, and we should have said an impossible one, seeing that at that moment she was the city and its population and its queen all rolled into one, so to speak. Queen-wasps, however, also on the American plan, ruled the word "impossible" out of their dictionary long ago. They "attempt the end, and never stand to doubt."

The queen came to rest on a bare patch of ground an front of a hole, and a black and hairy spider, with two hindlegs missing on the offside, spun round in the entrance of that hole to face her. He had not been noticeable until he moved.

She left him in a hurry, and thereafter resumed her endless searching along the hedge-bank. A dozen times she vanished into a hole, and, after a minute or so, came out again with the air of one dissatisfied. Half-a-dozen times she came out tail first, buzzing warnings and very angry, at the invitation of a bumble-bee queen, a big, hook-jawed, carnivorous beetle in shining mail, and so forth, but she never lost her head.

Finally, she came to a mole-hole that suited her. The other burrows had all turned out to be field-mouse holes, leading ultimately into a main tunnel that ran the whole length of the hedge apparently, and was a public way for all the little whiskered ones. But this tunnel, bored by the miner mole, ran nowhither, having caved in not far from the entrance, and was very sound of construction, with a nice dry slope. She selected a wide spot where the tunnel branched, each branch forming a *cul-de-sac*. Here she slew swiftly several suspicious-looking little tawny beetles and one field-cricket, who put up a rare good fight for it, found loafing about the place.

It pleased the queen that here, in this spot, she would found her a city. But first she must, as it were, take the latitude and the longitude of this her stronghold to be. She must know where her city was, must make absolutely dead sure, certain, of finding it again when she went out. Otherwise, if she lost it--well, there would be an end to it before it had begun, so to speak. For this purpose, therefore, she rose slowly, humming to herself some royal incantation--rose, upon a gradually widening corkscrew spiral, into the air.

She was, in point of fact, surveying the district round her capital to be, marking each point--bush, stone, grass-tuft, tree-trunk, flower-cluster, clod, branch, anything and everything, great and small--and jotting down in indelible memory fluid, upon whatever she kept for a brain, just precisely the position of every landmark. And as she rose her circles ever widened, so that at last her big compound eyes took in quite a big stretch of sunlit picture, to be photographed upon her memory, and there remain forevermore.

It took her some time, for it was some job; but once done, it was done for good.

Next, alighting with great hustle--now that the work was once begun--the queen ran into her tunnel, and made sure that nobody had "jumped her claim" in the interval. She found an ant, red and ravenous, taking too professional an interest in the place, and she abolished that ant with one nip; though, as you may be sure, the tiny insect fought like a bulldog.

Then she executed a shallow excavation upon the site of the future city itself, carrying each pellet of earth outside beyond the entrance. This also took time, though she worked at fever-pitch, almost with fury; but she managed to finish it, and fly away into the landscape in a remarkably short while, considering.

Here once again she appeared to be searching for something through the yellow sunshine and the falling blossom-petals--confetti from Spring's wedding. And presently she found it, or seemed to--an old gate, off its hinges. But the wood was rotting, and she was no fool. She knew her job--the job she had never done before, by the way--and after humming around it in a fretful, undecided sort of fashion for some while, she flew on. Apparently she was looking for wood, but not *any* wood. Cut wood appeared to be her desire, and that oak; at least, she put behind her a deal board lying half-overgrown, after one careful professional inspection.

Her way was through a perilous world, beset by a thousand foes, mostly in the nature of traps and lines and barbed-wire entanglements set by spiders. As a rule you didn't see these last at all--nor did she; but her yellow-and-black badge usually won her a way of respect--and hate--and she cut or struggled herself clear of such web-lines as her feelers failed to spot in time.

At last she found some real oak rails, and set to work upon them at once, planing with her sharp shear-jaws. A tiger-beetle, gaudy and hungry-eyed, sought to pounce upon her in this task. He was long-legged, and keen, and lean, and very swift; but she shot aloft just in time; and when she came down again, with a z-zzzzp, as quickly as she went up, sting first, he had wisely dodged into a cranny, where he defied her with open and jagged jaws.

Again getting to work, she planed off a pellet of good sound wood--it looked like a nail-scrape, the mark she made--and masticating it and moistening it with saliva, whirred back like a homing aeroplane to her city in the making.

There was a whir and a buzz as she passed through the portals of her main gate from the light of day, and she reappeared again, backing out, "looking daggers," as we say, and brandishing her poisoned dart--her sting, if you insist, on the end of her tail--in the air. But she still hung on to her pellet.

Presumably some unlucky visitor had called in her absence. More sounds of concentrated argument followed, and finally there fell out, rather than rushed out, a small and amazingly slender black wasp, one of those hermits who seem to consecrate their lives to lonely working for a family they never see. This unhappy one slid down the bank, curled up at the bottom, uncurled, curled up again, and--remained curled. Apparently *her* day's work was done, which comes of falling foul of the yellow flag.

Arrived inside, at her hallowed chamber, our queen carefully selected a rootlet in the roof--not just any old rootlet, mark you; never any "old" anything, you will notice, but a good, sound, well-found rootlet that you could hang five or six pounds' weight to; indeed, three rootlets before she had finished. To these rootlets she fastened--gummed would be a more correct word--her pellet of wasp-paper, in the form of a thin layer, and hurried away, singing, for more. This was, so to speak, the foundation-stone of the city, laid, be it noticed, not haphazard--our queen never did any business that way--but with mathematical regard as to what was to follow. In very fact, too, it was the foundation-stone of her city, only upside-down, though that is nothing. Wasps always do things that way, which is unlike ants, those other and greater city builders.

Back came the queen very soon with another load, and pasted that--thin--to the first layer, hurrying, bustling, humming a happy song continuously to herself. Then away again for more, and in the process to a lively battle with a robber-fly, who appeared set upon robbing her of her blood. It tried, like the beetle, to stalk her and pounce upon her back, what time she was planing out wood for paper-pulp; but her back wasn't there when it pounced, and her jaws were. It "waited on," hovering like a falcon, and twice as keen, and when she got to work again, dropped like a hurled lance-head, only to be met with jaws, wide and ready, as before.

It went away, watching from afar--far for an insect--from "the little speedwell's darling blue" upon the hedge-bank, and just as she was moistening the load, gathered ready to fly off, delivered its final ultimatum--a marvelously persistent murderer. This time it, or, rather, she, was received with the point--the poisoned point--and, turning like a spent lightning-flash to avoid it, found the queen hard on her heels, following all down the gay hedge-ditch, humming high, in nearly a shriek of rage. Finally, she turned, to do battle for her life, and the two, grappling, fell as shooting-stars fall, gleaming, athwart the sun, with a brrr-r like a fused wire, and finished the job, rolling over and over on the ground--rolling over and over among the stalks of bluebells, like the heavens "upraising from the earth."

It is written, however, that few in the insect world can withstand a queen of the yellow devils, and in a few seconds the wasp got up and flew home again, quite unperturbed. The robber-fly did not get up, and she was not quite unperturbed, but died as they die who are poisoned with formic acid, and very soon was still.

By the time the shadows crept across the entrance to the derelict mole-hole, warning the wasp back--for your true wasp is a worshiper of the sun--the queen had formed a disc of paper, and suspended there-from, in the middle, a stalk, also of paper, which widened out at its base, and became, as it were, the outlines of four six-sided cells. The cells were in the shape of a cross--that cross which you will always find at the foundation of the cities of the waspfolk, and, in a way, a sign or mark of their nationality--the cross in the market-square, so to speak, outwards from which the city grew.

The queen, satisfied apparently with her new city so far, hung up and went to sleep. When anything or anybody came to prospect for house lots, or edible victims, during the still, silent, silver night, she hummed very severely, like an electric fan, to let the intruders know who she was, and they mostly backed out again in a hurry. If they took a step nearer the hum rose an octave, and became very wicked, and that, so far as most of them were concerned, finished it.

Two, however, there were who would not take even that hint. One was a shrew-mouse, thirsting for blood, but who got poison instead, and next morning was found running about with his mouth somewhere concealed behind his ear, if one may be pardoned the expression, in consequence; and the other was a carnivorous beetle, in black, purple-shot armor, and armed with jaws toothed like lobsters' claws. The queen took some nasty scars from those same jaws before she got home with the poisoned point, a clean thrust 'twist breastplate and armlet, and the invader doubled up on the spot where he was, and had to be dragged out in the morning--not the dawning, for the sun had well stoked up before our wasp would have anything to do with him.

"A shrew-mouse, thirsting for blood, but who got poison instead"
—*Page 52*

[Illustration: "A shrew-mouse, thirsting for blood, but who got poison instead"]

She found the day already in full swing when she rose, buzzing, from her front-gate, late--for wasps hate early morning chill, like Red Indians--and, circling once, swung straight away. She jumped into full hustle right off, you see. She did not merely work; she superworked. Forced to short hours by her constitution, she had to make up for it in the time she got, and she did. She allowed nothing to stop her. If anything tried to, she mostly stopped it, for there was no compromise about this nation-builder; she reached her goal every time.

It was on this journey that a spotted fly-catcher, sitting on a gatepost, made a Euclid figure at her in midair as she passed. She had not power to fight the bird's beak, and her poison-dagger was useless here; nor do fly-catchers often miss. This, however, was an occasion when one of them did--by an eighth of an inch--and only some electric-spark-like dodging on the part of the insect in the air made even that one miss possible. It was so quick, you could not see what happened.

That day the cross of cells in this budding city was developed further, and a low wall built round each cell. Moreover, more cells were built, always taking the cross as the center of all things--six-sided cells, with a low, incomplete wall, or, rather, parapet, partitioning each off, to the number of about twenty-four cells in all. Each cell was closed, of course, at the top, the top being its floor, and open at the bottom, the bottom being, if I may so put it, the top; for, as has already been said, wasp cities are built upside-down, and everybody walks and hangs on his head, being so fitted for the purpose. If you don't hang, you tumble straight down into the scooped-out cavity below; but nobody ever does that till he dies, for that cavity is at once the cemetery and the refuse-heap and the dust-bin of the city, a haunt of tiny ghouls--beetle, spider, and fly ghouls--and other loathsome horrors, the scavengers, hyenas, vultures, and jackals of the wasp world.

Now, after making the first cell, or, rather, the part cell, with its low parapet, the queen laid an egg--it was very minute, that egg--inside the cell, gumming it against the top, on the angle nearest the center of the city. It had to be cemented there; otherwise it would have fallen out.

In the next cell she laid an egg, too, cementing it up to the top in the same manner--always in the angle nearest the center of the city--and in the next another egg, and so on, up to the twenty-four or so. It is a little doubtful precisely how long she took over the process, because, for one thing, she made so many journeys backwards and forwards to get wood-pulp from the rails for paper manufacture--she used paper for everything; and, for another thing, she began to roof over the whole affair with a hanging umbrella made of layers of the finest paper that you ever did see--much finer than that made by the ordinary common or garden worker-wasp of the jam-pots and the stewed-fruit dish, for was she not a queen, and therefore not common in anything she did?--and it became, in consequence, rather hard to see what she really was "at." Most of the time that the sky remained cloudy she used up at this job, and also when there was a shower of rain, for she hated rain and all shadow and darkness.

Her purpose, in regard to this paper roofing, was to keep out any possible dripping that might come through the earth roof in wet weather, and to store up and multiply the heat from her body. Terrific heat, to be sure; nevertheless important in the scheme of things. When all was completed, this city, this mighty kingdom, measured about one and a half inches round.

When all was completed, also, the wasp flew out for a drink and a feed. But first she cleaned. The most fastidious cat was a grimy tramp in comparison to her in habits, and in all her spare time--goodness alone knows how she squeezed in any spare time at all during those hustling days!--her first, and generally her last, act was to clean. She could not afford dirt. To be dirty, with her, was to die even more quickly than she would, anyway; for, you see, she did not breathe through her mouth, but all over herself, so to speak--through her armor, or hair-like tubes in that same.

From bluebell to cowslip and lily she picked her way, sipping honey and humming a wicked little hum through her teeth, as it were, and on to where

> Daisies pied, and violets blue,
> And cuckoo buds of yellow hue,
> And lady-smocks all silver white,
> Do paint the meadows with delight.

Now she toyed with a yellow oxlip, now paused at a purple lungwort; but most she went into the garden, and hovered, still as a humming-bird, among the rose-leaves and branches, especially those growing against the sun-bathed old wooden porch, and for so long that one wondered what she was doing there. She was licking up the "honey-dew," which, translated, is the juice exuded by the plant-lice or "green-fly," which swarmed all over the rose-trees. This "honey-dew" was sweet, and in great demand among such insects as had tastes that way; in fact, the enterprising ants--who are always a decade ahead of everybody else--were, in one place, building mud sheds over the said herds of plant-lice to prevent their precious "honey-dew" being exploited by others.

Thus a week passed, the queen fussing daily about her embryo city, adding paper covering here, strengthening a wall there, warning off an inquisitive insect somewhere else, and adding her heat to the natural stuffiness of the place, though one would scarcely have thought she could have made much difference. At times, too, in the hot sun, she appeared here or there outside, drinking honey from some flower, or sipping "honey-dew," much to the ants' disgust and anger.

Then, at the end of the week, the first egg hatched out within the city, and, frankly, what came forth was not lovely. It was a legless grub, fat, presumably blind, and helpless; and it would have fallen head downwards out of the cell, as it hatched, if it had not had the sense to hook its tail into its own egg-shell, which in turn, as we know, was already fastened to the top of the cell. But it had jaws, and in addition, apparently, an appetite to use them.

Whether the queen loved it, her first baby, was hard to tell. Did she, indeed, ever love anything? She certainly did her duty by it; but what was the use of setting up to be a queen, anyway, if she could not do that? And, moreover, you've got to do your duty in the wild. There's no profit in monkeying with Nature, as is possible with civilization, for the penalty thereof is death.

Wherefore did our queen, after making quite sure that the sack-like atom with a mouth, hanging upside-down in the cell, and wriggling like anything to show its hunger, was alive, sound, and quite all there, quit home in a hurry, and with a loud buzz, in search of rations. But there was a change in her manner from that adopted when looking for food for herself, and for good reason. *Then* her object had been honey; now it was--scalps!

From force of habit, more perhaps than from force of reasoning, she flew to the rose-trees, and there fixed in her shear jaws not more than two of the helpless, fool, unarmed, soft, juicy green-fly, which are really no more, if one may so put it, than living, infinitesimal "white" grapes. That she was challenged by a sentry ant--about as big to her as a bulldog to us--that the sentry gave the alarm, that the guard turned out from one of the ants' "cowsheds" over some of the green-fly, and that she went away in a hurry, with half-a-dozen furious ants on their hindlegs, trying to get hold of her retiring feet with their jaws, was a matter treated by her with insolent unconcern.

She had got her scalps, and winging home in a hurry to her baby, fed it upon green-fly. The baby did not feed nicely, and the picture of the glistening, corsleted devil queen-mother, with her lugubrious, mask-like face, and the wriggling, hanging sack babe, and the luckless, fool, helpless green-fly between them, was not a pretty one. Here maternity was not a Sunday-sermon subject, yet it was maternity all the same.

By this time other eggs in other cells were splitting, and giving out legless grub horrors, as seeds that give forth plants, each wriggling mummy taking care to hook itself up to its shell by the tail at once, lest it perish. And the queen's work from that moment really began. Till then she had only tinkered at it, apparently. Now she got going "real some," and--well, all the insect world outside knew it. The terror of the yellow flag spread.

Upon an hour she would appear, dropping, hawk-like and terrible, out of the sun-glare, and neatly pick up a soft and juicy caterpillar from a cabbage-stalk. Upon another hour she would be discovered, feet tucked up and wary, darting, like an iridescent gleam, around the angry ants, among the green-fly on the rose-bushes. The drowsy hum of the kettle on the kitchen fire, and the steady, low hum of the house-fly dance in the middle of the room, would be answered in the long, hot afternoons by her wicked warning drone as she came sailing in at the open window, like the insolent pirate that she was, to go out again a minute later with a helpless fly between her jaws. The first heat of the sun, drinking up the dew, would discover her sailing forth to war; his full, sizzling rays would reveal her waging violent warfare with the bluebottle flies over some carcass; into his amber light of the noon her yellow flag would suddenly rise from out the cool shade of the larder, where she had been carving meat, and "when the sun mended his twisted copper nets," he would flash in bronze from her glistening cuirass as she droned by high over some wriggling grub, caterpillar, or palsied fly fast locked in her jaws--and all for her young, all for her couple of dozen legless horrors, hanging by their tails, each in its narrow cell, in darkness and in dead silence, in the embryo city under the secret earth.

Time was when these same grubs grew so fat and big that they no longer hung, but became fast wedged in their dormitories; time when the queen had to set to and extend downwards the wall of each cell lest the growing inmates bulge over, and, obsessed with their ravening hunger, incontinently eat each other; and time at last when, one after the other, each grub, having grown out of more than one suit of clothes and donned new ones, cast its skin for the last time, refused all further food, spun a cocoon of silk with a dome-shaped silken floor to each cell, and for a period retired from the prying eyes of the world, even of its own mother, into the sacred sanctuary of the chrysalis state. Then the queen's labor lightened a little for a period, so that you could again see her at spare moments sucking nectar from the flowers for herself, robbing the jam-dish, or lapping up the "honey-dew" of the green-fly.

Finally came *the* day. It dawned all right, and there was nothing about it to show that it was going to be different from any other fine day; yet, as soon as the wasp woke up, she knew that, for her, it was the day of Fate.

A very cursory inspection of the budding city showed at once that during the night things had been happening and changes taking place. The domed floors of several of the cells were palpitating with life from within, and there were sounds of the gnawing and tearing of the silken screens.

The queen became greatly excited, and began to hum and dance a little step-dance to herself, all alone in the darkness among the cells, as she saw her triumph evolving before her eyes. And, almost as if the hum had called it, there rushed at her, out of the blackness across the comb, a--a thing.

She knew by instinct that it was an enemy. Indeed, it could not well be anything else, but it fought like a black devil.

It was, in point of fact, a mole-cricket, a creature just like its namesake, if an insect can be said to resemble an animal, only that its jaws were like unto the jaws of a lobster. It was a fearsome apparition, and very much larger even than the queen. The good God alone knoweth why it had chosen that moment and place to run apparently amok.

But, if the mole-cricket ran amok, the queen-wasp went berserk. It was a thing unthinkable that in that moment of triumph she, and the awakening city with her, should be cut off--unthinkable and impossible, unthinkable and maddening. Therefore she fought as few wasps have probably fought before or since, and they are pretty expert exponents, and scarcely backward ones, of warfare.

The battle that followed was awful.

Almost at the start the two insects, grappling, fell headlong to the excavation the queen had made below the city, and there, rolling over and over, continued the struggle in the dark among the refuse, the queen eternally feeling with her poison-dagger for a space to drive home her death-blow between the other's smooth, shining armor-plates; the cricket eternally endeavoring to behead the queen between its awful jaws.

It was a fight to the death, as most insect duels are, and it could not last long. It was too tense, too fiendish, too shockingly wicked for that.

Suddenly the queen's body shot out like a spring. The opening she had been feeling for had appeared, and she had driven her death-blow home. At the same instant, with a supreme effort, she bent double and shot herself free, the last convulsive, shearing crush of her foe's laws clashing to so close above her head that they actually caught in their death-grip, and held, till she pulled them out by the roots, two bristles of her neck.

And then--well, then the queen hurried back up to her city, just in time to help out of its cell the first of her children--and citizens at last--the first limp, clambering, damp, newly painted, freshly bedecked young worker-wasp, perfect from feeler to sting, from wing to claw.

Quickly they broke out now from the cocoons, and the queen bustled from one to the other, assisting, cleaning, encouraging; for it is a tricky job for an insect to come out of its chrysalis-case. The queen's work, however, was really done; for, though for a day or two, till their cuirasses and wings hardened, these new young worker-wasps only did light labor, acting as nurses to the others that were following, and so on, they quickly took upon their own shoulders the whole of the work of the city: the nursing and feeding of the young, the hunting, the building, the scavenging, and the waiting upon and feeding the queen-mother herself completely, so that she should henceforth labor not, nor fight, nor waste herself in the chase, but should keep at home and lay countless eggs, and eggs, and always nothing but eggs, for the workers to rear for the benefit of the State.

IV

NINE POINTS OF THE LAW

Sharp's the word with her.--SWIFT.

Some people never know when they are well off. It is a complaint which afflicts cats, you may have noticed, and gets them into much trouble that their contemptuous temper might otherwise leave them free from. The silver tabby would have done better if she had remained asleep upon Miss Somebody's arm-chair, instead of squatting, still as marble, out in a damp field on a damp night, watching a rabbits' "stop"--which is vernacular for a bunnies' nursery--and thinking how nice raw, pink baby-rabbit would taste if she got the chance to sample it. She didn't. At least, she hadn't for an hour and a half; but, then, what's an hour and a half to a cat? Apparently the silver tabby could wait, just like that, utterly inert, till the crack of doom--or dawn.

Mind you, she was not alone. She had company. One always has in the wild at night, or nearly always. You couldn't see that company, but I don't know whether the silver tabby could. Who can tell how much a cat sees, anyway? Nor do I think the company could see her, she being still, and wild eyes not being good at picking out the still form. Neither could they hear her, for she said nothing; neither did she purr. They must have smelt her, though. Anyway, she seemed to be a little island in the mist--the faint, faint, ethereal dew-mist--where nobody walked. You could hear them--a rustle here, a squeak there, a thud somewhere else, a displaced leaf, a cracked twig--this only once--a drumming, a patter, a sniff, a snuffle, a sigh; but they all passed by on the other side, so to say, and gave the silver tabby room to think. Apparently cats are not considered good company in the wild; lonely creatures, they are best left alone.

No mother-rabbit came to the "stop"--which the cat knew to be there--to feed her babies, which the cat, thanks to her nose, knew to be there, too. No baby-rabbits came out to be fed--or to feed the cat either. "Stops" are secrets, kept from the rest of rabbitdom by the wise mothers, and, they hope, from other inquisitive people also. The little short holes in the middle of the field are plugged up by the old does with grass and fur when they are away, which is pretty often.

Then the silver tabby heard a thump come out of the night--a thud, hollow, resounding, and noticeable. It was repeated after an interval, and again repeated. There was a certain note of insistence about it--like a signal. And if the cat had been a wild creature she would have thrown up the sponge, or gone away, and returned secretly later, or, anyway, not persisted in crouching there; for those thuds were a signal, and they meant that the game was up. In other words, some wily old mother circling the approach, or some wandering back-eddy of wind, had given the cat away; she had been scented, and rabbit after rabbit, squatting invisible in the night, was thumping the ground with its feet to say so and warn all off whom it might concern. The silver tabby, however, neither wild nor satisfied to be tame, did not know. She sat on, and in doing so wondered, perhaps, at the scarcity of rabbits thereabouts.

She sat, or hunched, or crouched, or couched, or whatever you call that precise position of cats, which is neither lying down nor sitting up, for some time longer--for another twenty minutes, to be precise; and all the while the thuds of mystery serenaded her from nowhere in particular out of the dark--and from down-wind.

Then she must have come to the conclusion that she was being made a fool of, for she got up, stretched herself lazily, with arched back and bared claws, and yawned a bored feline yawn. And even as she did so she was aware of a sudden final flourish of thuds, and then dead-silence. Next moment, in that same dead-silence, she distinctly heard something coming towards her, and that something was taking no pains to conceal the fact.

Now, in the wild it is not the custom to go towards anything and take no pains to conceal the fact. The unhealth of such a procedure is swiftly borne in upon such rash ones as make the experiment, and they seldom live long enough to pass their folly on. Only the mighty can afford not to walk circumspectly, and they are very few, and, with man about, even they have learnt wisdom. That is why the wild is so guarded, and why self-effacement becomes almost a religion therein.

Even the cat knew this, I fancy. Anyway, she looked surprised as she crouched again, and quickly.

Now, of all the wild-people, probably one of the most brazen is the pig; it is also one of the bravest. I mean, the wild pig. And it would seem that he, or she, who came that way was a pig, only a precious little one. You know the ways of a pig? How you can hear him coming long before he comes; how he must snuffle, and grunt, and poke dead leaves, and snort, and tread on things, and snore. Very good. So it was here; and these things did this new-comer, who approached through the mist--only all in a dwarfed way, as if they were done by a tiny grown-up pig. Its gruntings were almost to itself; its snortings, snorings, and sniffings quite small; and its snorts little miniature ones. Only, in the profound silence of the night, and in comparison with the furtive noises of all the rest of the night-wild, they sounded quite loud.

The cat, as she crouched, passed from supercilious surprise to amazement. You could tell that by the roundness of her eyes. She had no knowledge of pigs, and had never met any of the wild-folk gone mad; yet it seemed that one must have done so now, and that one--to her growing uneasiness--was coming straight towards her. I fancy that in that moment she thought of the warm fire, the singing kettle, the saucer of milk, and Miss Somebody's best arm-chair.

The thing, whatever it was, came straight on in a more or less zigzag line, till the cat could make it out dimly in the moonlight, a blotched, roughly egg-shaped form, less than a foot long, so low to the ground that it appeared to be running on wheels, and covered all over with prickles, like a Rugby ball into which tin tacks had been driven head first, the sharp ends pointing outwards and backwards. Its head was the small end, and much lower than its back. Its eyes, little and pig-like, set in a black cowl, gleamed red in the tired moonlight; and its face was the face of a pig, nothing else--just pure pig; insolent, cunning, vulgar, and blatant. Occasionally men name a wild beast correctly, and this little beast could only have one name--hedgehog: It was obvious on the face of it.

But the cat, being a cat and an aristocrat, knew, as has been said, nothing about pigs, real or only so called. She had killed a shrew once, and spat it out for tasting abominably and smelling worse; and shrews are cousins of the hedgehogs, of the same great clan, Insectivora--far removed from the pigs, really--and that is the nearest she had got.

She had never heard of hedgehogs, and never, never met a beast that walked through the wild as if he owned it. And, more, he expected her to get out of his way, which she did with feline and concentrated remarks; and he--by the whiskers and talons!--the fool exposed his back--turned his back openly, a thing no wild beast in its senses would do, unless running away. And that, for a cat who had waited close on two hours for baby business that didn't turn up, had got most unfashionably drenched, and had, moreover, in her time, tackled more than one grown-up rabbit, which was considerably larger than any hedgehog--that, I say, was, for the silver tabby, too much.

She sprang. Rather, she executed two bounds, and somewhat unexpectedly found herself on top of the hedgehog. I say "unexpectedly," because she had hitherto bounded upon wild-folk who contrived mostly not to be there. This one contrived nothing, except to stop still. And the cat executed a third bound--*off* the hedgehog, and rather more violently and more quickly than the first two. Also, she spat.

When she had got over the intense pain--and cats feel pain badly--of sharp spines digging into her soft and tender forefoot-pads, she stopped, about two yards away, and glared at the hedgehog as if he had played off a foul upon her, and she was surprised to see that he was no longer egg-shaped, but rolled up into himself like a ball, so to speak, and utterly quiescent. (I wonder if she remembered the little wood-lice that she had so often amused herself playing with in idle hours. They rolled themselves up just like that. Perhaps she thought she'd come upon the Colossus of all the wood-lice.) Anyway, after she had spat off at him all the vile remarks she could think of for the moment, without producing any more reply than she would get from the average stone, she came back, drawn with curiosity as by strings.

The hedgehog did not move; there was no need. It was for the cat to make the next move--if she chose. He did not care. All things were one to him, and all the views which he presented to the world were points, a *cheval-de-frise,* a coiled ball of barbed wire, a living Gibraltar, what you will, but, anyway, practically impregnable; and the beggar knew it. "He who believeth doth not make haste"--that seemed to be his motto, and he had, by the same token, a fine facility for withstanding a siege.

He felt the cat, that cat who did not know hedgehogs, pat him tentatively. Then he heard her swearing softly and tensely at the painful result. She did not pat again--at least, only once, and, in spite of care, that hurt her worse than ever. Then she began growling, low and beastily--for all the cat tribe have a horrible growl; you may have noticed it. Perhaps the hedgehog smiled. I don't know. He knew that growl, anyhow; had heard it before--the anger of utter exasperation. He was an exasperating brute, too, for he never said anything, only shut himself up, and let others do the arguing, if they were fools enough to do so.

Suddenly he heard the growl stop. Followed a tense pause, during which he tightened his back-muscles under his spines, and tucked himself in, to meet any coming shock, more tightly than ever. Followed the pause a short warning hiss, jerked out almost in fright, it seemed--that cat's hiss that is only a bluff, and meant to imitate a snake--a sudden explosion of snarls, and a thud. A fractional silence, then a perfect boil-over of snarls, and thud upon thud.

Now, our friend hedgehog was an old hand, and he had heard many and curious sounds take place outside himself, so to speak; but, all the same, he was just tickled to death to know what, in claws and whiskers, was happening out there in the leering moonlight now; so much so, indeed, that at last he risked it, and took a furtive peep out of a chink in himself, as it were. And what he saw might have amazed him, if he had not been a hedgehog and scarcely ever amazed at anything. He just got a snapshot view of the cat's fine ringed tail whirling round and round as she balanced herself on the swerve, vanishing into the ghostly moonlight haze of the night; and in front of him, close beside him, squatting, stare-eyed and phlegmatic, he saw the form of a big, gaunt, old doe-rabbit. And I think he knew what had happened. He seemed to, anyway, and remained rolled up.

Rabbits are thoughtless, headstrong, headlong, hopeless, helpless cowards as a race and a rule. "The heart of a rabbit," they say in France, speaking of a coward. But all races and rules have exceptions. Occasionally the exceptions are old buck-rabbits, who know a thing or two; but more often they are old doe-rabbits with young. And, mark you, from the point of view of those wild-folk, there may be easier rough handfuls to tackle than old doe-rabbits with young. This one had simply streaked out of the night from nowhere--and behind--and knocked the cat flying before she knew. Then, ere ever the feline could gather her wits, the old doe had descended upon her with an avalanche of blows--punches they were with the forefeet, all over the head and the nose, where a cat hates to be hit--and all so swiftly, so irresistibly, that that cat had never been given a chance to consider before she was stampeded into the night. It was the silver tabby's first experience of Mrs. Rabbit doing the devoted-mother act, and, by the look of her--tail only--and the speed at which she was going, it appeared most likely that it would be her last.

"This one had simply streaked out of the night from nowhere"—*Page 69*

[Illustration: "This one had simply streaked out of the night from nowhere"]

Meanwhile the old doe-rabbit sat there in the moonlight as immovable and impassive as a Buddha, and the hedgehog, peering at her, guessed that the time to unroll was not yet. He knew that it would hurt any one to attack him; the cat knew it; all rabbits in their senses knew it; but was that mother-rabbit in her senses? He concluded to lie low and remain a fortress, therefore.

Then, after waiting about five minutes, as if she knew that cats sometimes steal back, the old doe-rabbit came to a "stop" quite close to the hedgehog, and went in. She remained there some time, during which a fox came by and sniffed at the hedgehog, but was quite wise as to the foolishness of doing more; and a deadly, curved-backed, flat-headed little murderer of a stoat galloped by, and sniffed too, but was no bigger fool than the fox, and went his way.

Both missed the "stop" by about two yards, though I don't know what would have happened if they had found it. Digging and death in the former case, and battle and blood in the latter, perhaps. But no matter, they passed on their unlawful occasions; and half-an-hour after the going of the stoat the old doe-rabbit came out, and dissolved into the moon-haze.

Then the hedgehog came out, too--of himself, and--well, dissolved into the "stop."

What happened in there it was too dark to see, but not to hear; and what one could hear was--pitiful. He was there some time, for your hedgehog rarely hurries; and when he came out again, his little pig's eyes gleaming red under their spined cowl, it was with the same snuffling, softly grunting deliberation with which he had gone in; but the pale moon, that showed the gleam in his eyes, showed also blood on his snout, and on the bristles of his forefeet, blood.

Then, slowly, snorting, sniffing very audibly--as loud as a big dog often does--grunting softly in an undertone, as if talking to himself, he departed, rustling through the grass, leaving an irregular winding track behind in the dew and the gossamer, as he searched, eternally searched, for food.

The hedgehog moved through the night as if he owned it and had no fear of anything on earth; but many, it would seem, had cause to fear him. He turned and snorted, and snatched up a slug. Three very quick and suggestive--quite audible--scrunches, and it was gone. He described a half-circle, sniffing very loudly, and chopped up a grub. He paused for a fraction to nose out a beetle, and disposed of it with the same quick three or four chopping scrunches. (It sounded rather like a child eating toast-crusts.) He continued, always wandering devious, always very busy and ant-like, always snorting loudly; grabbed another beetle, and then a worm--all by scent, apparently--and reached the hedge-ditch, where, in the pitch-darkness, he could still be heard snorting and scrunching hapless insects, slugs, and worms at scarcely more than one-minute intervals. And he never stopped. He seemed to have been appointed by Nature as a sort of machine, a spiked "tank," to sniff tirelessly about, reducing the surplus population of pests, as if he were under a curse--as, indeed, the whole of the great order of little beasts to which he belonged, the Insectivora, are--which, afflicting him with an insatiable hunger, drove him everlastingly to hunt blindly through the night for gastronomic horrors, and to eat 'em. Anyway, he did it, and in doing it seemed to make himself worthy of the everlasting thanks and protection of the people who owned that land--thanks which to date he had never received.

Strange to say, he never stopped of his own free-will, though he was stopped: once when he walked up to a man kneeling--and he was a poacher--and did not see him till, if I may so put it, the man coughed, when he ran like winkle into the hedge, and promptly became a ball for ten minutes; and once when he came upon a low, long, sinister, big, and grunting shadow, which again, if I be allowed the term, he did not see, though quite close, till he heard it grunt, when he instantly jerked himself into a ball on the spot and in the open. In both cases it seemed, on the face of it, more as if he had scented, rather than had either seen or heard, the dangers, and in both cases he had come within two yards of them--though they were not hidden--before scenting, seeing, or hearing them, whichever he did do.

Now, books and men have said that friend hedgehog fears only two things: gypsies and badgers--who eat him. I should not be surprised at anything the "gyp" did; nor, to this day, can we stake much on our knowledge of the secret badger; but this badger, at any rate, seemed to know nothing of books and men. He was delving for roots when the hedgehog cast up out of the night and jumped him to "attention" by his loud sniffs--much like a big dog's, I said. Thereafter, however, when our prickly friend was represented as a ball only, and was as silent as the grave, the badger took no further notice of him, beyond keeping one eye--the weather eye--upon him, and treating him to a low growl, or curse, truly, from time to time.

The hedgehog, however, once there, did not seem keen upon unrolling and exposing himself till the badger had gone, which it did finally, vanishing so suddenly and unexpectedly into the dark as almost to seem to have been a ghost. And after some minutes the hedgehog straightened out, and ate his way--one can call it nothing else--to the hedge. Here he came upon a wounded mouse, complaining into the night in a little, thin voice, because its back was broken, and it could not return to its hole. It was a harvest mouse, rejoicing in the enormous weight of 4.7 grains and a length of 57 mm., but with as much love of life and fear of death as an elephant. Heaven knows what had smitten it! Perhaps it was one of the very few who just escape the owl, or who foil that scientific death, the weasel, at the last moment--but no matter. The result was the same--death, anyway.

The hedgehog saw its eyes shining like stars in a little jet of moonlight, and I fear the hedgehog slew far less adroitly than the owl, and not nearly so scientifically as the weasel; but he slew, none the less, and he did that which he did.

From thence we find our hedgehog, still wandering devious, but with always a direction, just as an ant has, heading his way down-ditch to a farm, and all the way he ate--beetles mostly, but with slugs and worms thrown in.

Now, those of the wild-folk who approach the farm, even by night, do so with their life in their paws, and most of them know it. Far, far safer would it be to remain in wood or field-hedge, gorse-patch or growing crop. Yet they go, like the adventurers of old.

First of all, if he approached by ditch, before getting to the farm proper, the hedgehog knew that he must pass the entrenchments of the rat-folk, and that alone was enough to put off many, for the rat-folk are no longer strictly wild, and, wild or tame, are hated with that cordiality that only fear can impose. I don't know that our hedgehog was given to fearing anything very much. He came of a brave race, and one cursed, moreover, with a vile, quick temper, more than likely to squash in its incipient stage any fear that might threaten to exist; but he did most emphatically detest rats, except to eat them--a compliment which the rats would have returned, if they had got a chance.

As a matter of fact, it is unlikely that Prickles--for such was the name of our hedgehog--would have gone that risky way, traveled so unhealthily far, left his more or less--mostly less--safe home wood at all, had it not been that it is sometimes with hedgehogs as it is with men--in the warm seasons--their fancy turns to thoughts of love. Prickles's fancy had so turned, not lightly, for he was of an ancient and antediluvian race, heavy in thought, but certainly to love. And love, I want you to realize, in the wild, or anywhere else, for the matter of that, is the very devil. "Unite and multiply; there is no other law or aim than love," one great savant despairingly assorts is Nature's cry, and adds that she mutters to herself under her breath, "and exist afterwards if you can. That is no concern of mine."

To be precise, Prickles, who did more business with his nose than all the rest of his organs put together, was following a love-trail. A lady hedgehog, a flapper undoubtedly, and beautiful--all loves are beautiful in imagination--had passed that way. Why *that* unhealthful way, Heaven knew; but, allowing for the capriciousness of the sex, and mad because in love, Prickles followed, slowly, deliberately, heavily, as befitted one descended from one of the oldest races on earth.

The air was heavy with the scent of may and of honeysuckle, and his way was a green-gold--silver where the moon cascaded down the hedge--and blue-black bridal-path, arched with scented swords, strewn with pink and rose and cream and white confetti of blossom. But he only saw and smelt one thing, and that, those who have known hedgehogs intimately will agree, is not like unto the scent of any blossom.

Prickles was ruminating anciently upon these things, possibly, and others, as he came down the trench--ditch, I mean--when the cry smote him. It smote everything--the filtered silence of the wonderful, tranquil night, the pale moon half-light, the furtive rustling shadows that stopped rustling, the wonderful breathing pulse of growing vegetation. And Prickles stopped as abruptly as if it had smitten him on his nose, too. He heard *that*, at any rate, whatever might have been hinted about the value of his ears elsewhere.

There was no doubt about that cry, no possible shadow of doubt whatever--it was a cry of extreme distress, a final, despairing S.O.S., flung out to the night in the frantic hope that one of the same species would hear and help.

Several night-foraging wild-folk have S.O.S. signals of their own, but none like this. It was not a rabbit's cry, for bunny's signal is thin and child-like; nor a hare's, for puss's last scream is like bunny's, only more so; nor a stoat's, for that is instinct with anger as well as pain; nor a cat's, for that thrills with hate; nor an owl's, for that is ghostly; nor a fox's, for Reynard is dumb then; nor a rat's, for that is gibbering and devilish; nor a mouse's, for that is weak and helpless. Then what? And why had it touched up Prickles as if with a live wire? It was perhaps the rarest S.O.S. signal of all heard in the wild, or one of the rarest, the peculiar, high, chattering, pig-like, savage tremolo of a hedgehog booked for some extra deathly form of death. And Prickles--naturally he knew it.

It came from straightaway down the ditch; from ahead, where Prickles had been heading for; from the farm, and Heaven know what it portended! Perhaps, too, Prickles could tell a lady hedgehog's S.O.S. from that of a gentleman of the same breed; or, perhaps--but how do I know? He certainly acted that way.

Prickles waited the one-fifth part of an instant, to listen and locate. Then he got going, and provided one astonishment. Till then he had seemed slow as the times he had descended from--like a rhinoceros. But, like a rhino, he proved that he could shift some when hustled. He did. It looked like suddenly releasing a clockwork toy wound up to breaking-point. His short legs gave this impression, and his next-to-no-neck, giving him a look of rigidity, assisted it. He did not run so much as rush, and his spines and bristles, coming low on either side in an overhang, so to speak, like an armored car, made him rustle and scuffle tremendously. Three rabbits doing the same act, or five cats, could scarce have made more row than he did.

It was not, however, so much the fact that Prickles had gone that was so noticeable as the fact that he had *arrived*. His arrival seemed to follow his going as one slide follows another on a screen. One would never have believed such quickness of him; nor, as a matter of fact, do I think he would have believed it of himself; but--well, love is a mighty power, and makes folks do some strange things.

What he found was two ditch-banks, pock-marked with the untidy dug-outs of the rat-people, smelling ratty, and looking worse, one original ray of moonlight lighting the beaten ditch between. In the moonlight one young female hedgehog, who may have been pretty by hedgehog standards, but was now pretty by none, and five rats, frankly beastly, very busy indeed with that same hedgehog. They must have caught that young lady of the spikes "napping"--a rare thing. Yet, allowing for the fact that she was in love--with love and nothing else, so far--and careless, or allowing that she may have mistaken the unclean ones momentarily, she may have given them one brief half-instant. And it doesn't do to give a rat even the half of a half-instant. If you do, he has got you, or you haven't got him.

Apparently they had pretty well got her before she could quite roll up, and in a half-rolled-up condition she was doing her best to meet the jabs of five pairs of gnawing, cold-chisel, incisor, yellow-rats' teeth at once. To time, apparently, she had not been successful in the attempt--you could see the dark stains of blood glisten in the moonlight, and the end was certain, on the face of it.

Prickles, however, was a new factor that had got to short-circuit that end, and Prickles didn't wait to meditate prehistorically *that* time. He came. He came full tilt into the midst of the melee like--well, like a clockwork toy still, that couldn't stop. Only he did stop, against the biggest rat of all, ducking his head, and jerking forward his shoulder-muscles, and spines, with a sort of a thrust over his head, and a noise like a pair of expiring bellows; and the prickles hit home.

That rat removed about one foot in one bound in one-fortieth of a second, and he let rip one squeal in the process that sent away every other rat into the nearest available hole as if it had been fired there from a spring. Then the lady hedgehog took the Heavensent opportunity to complete her rolling-up completely, and Prickles took his own created opportunity to roll up almost more completely, and--well, they were rolled up into two balls, you see, and there is nothing more to be said about them. The rats did that, but it was all they did, except hurt their noses presently, and delicate, pink, hand-like forepaws, and make 'em bleed on prickles. They were very angry indeed, those unspeakable ones--very angry; but it didn't make any difference to the hedgehogs. They were there; they were rolled up; they were together. What *could* make any difference after *that*? And at last, when the rats gave them up as a very bad job, they went away *together*, and that's all there is to say. *Together* clinches it, you understand.

V

PHARAOH

I

Upon a day Hawkley came to the district, and took up his abode in a cottage of four rooms. He "did" for himself. Every housekeeper will know what "did" for himself means. But he did for himself in another way also. He came to read up for an exam. He told everybody this, which was one reason why he would be seen at ungodly hours, when no one was about, going to and from lonely spots, with a pair of blue glasses on his nose, a book under one arm, and a walking-stick with a silver band and a tassel--he was always careful to display the silver band and the tassel--under the other.

Then Nemesis descended upon him.

He was caught by Colonel Lymington's head-keeper on Colonel Lymington's most strictly preserved wild-bird sanctuary, shooting certain rare birds--many rare birds. Now, the colonel prided himself on his sanctuary, and upon the number of rare birds he had living therein, and the colonel was wroth. Hawkley had, in fact, ruined the sanctuary, and taken or slain pretty well every other bird worth having in the place, so that five years would not make good the harm he had done. Moreover, it was shown in the evidence that Hawkley had been able to accomplish his work by aid of a folding pocket-rifle with a silencer on, and his cat--especially the cat, whose name was Pharaoh.

No words of the keeper's could be found sufficiently to revile that cat. Indeed, the head-keeper went speechless, and nearly had epilepsy, in trying to describe it to the Court, and if it had done only one-half the things that the keeper asserted, it must have been a very remarkable beast indeed; the magistrate said so. In consequence Hawkley got rather heavily fined, and went. He went more quickly than was expected, because the police got a telephone message from the police of another district--several other districts, I think--to say that he was "wanted" for precisely the same game there: and Hawkley must have expected this, for he walked out of the court with a grin on his face, and was no more seen.

So quickly did he go that he had no time to take the cat. He left it at home in the cottage--which shows that he must have been badly scared, for such a cat must have been worth a lot to a collector's agent, such as Hawkley was. But perhaps he left it by way of revenge. I do not know.

Anyway, there it was in his cottage, asleep on the sofa before the fire--just as Hawkley, at the invitation of the authorities, had left it that morning.

It was about five o'clock in the afternoon when the cat, Pharaoh, woke up, and transformed himself instantly from deep sleep to strained alertness, in that way which is peculiar to the children of the wild, but has been lost by their domestic degenerates. The sun was shining full in at the little diamond-paned window. The window was open, and a late fly of metallic hue was shooting about with a pinging noise, like the twang of some instrumental string. But neither fly, nor sun, nor the tick of the little clock on the mantelpiece had awakened the cat. It was the click of the little front-gate latch.

The cat--the pupils of his eyes like vertical slits in green-yellow stone--gave one quick look at, and through, the open window. He had the impression, framed in the window, of a bobbing, black, "square" bowler hat--not often seen these days--and a red face with small eyes, and a sticking-out beard of aggressiveness. This was no Hawkley. The cat knew it, as he knew, probably, the alien tread. Hawkley had a white, clean-shaven face, and big eyes--the eyes that an animal may love and trust. Possibly the cat knew even the profession of him who came that way so softly and alone in the still afternoon. Anyway, he acted as if he did.

Like a snake, and with rather less noise, Pharaoh slid off the sofa and to the door leading into the scullery. For a moment he stopped, looking back over his shoulder, one paw uplifted, body drooping on bent legs, inscrutable, fierce eyes staring. Then he was gone.

I don't know how he went. He just seemed to fade out in the frame of the doorway and into the shadowed coolth of the scullery like a dissolving picture.

A pause followed, while the little clock on the mantelpiece ticked hurriedly, as if anxious to get on and pass over an awkward moment.

Came then the click of the front-door latch, the flinging open of the door wide, the bar-like gleam of hastily raised gun-barrels in the new flood of light, and--silence. Only the one or two late flies "pinged," while the little clock fairly raced.

The tall, uncompromising figure of the head-keeper was standing in the doorway, with a double-barreled 12-bore gun half-raised.

He stood there a moment with his dog, bent a little, peering in. He had come to find "that there pesky cat." And in this, perhaps, he showed more sense than most people gave him credit for. Apparently, he had seen enough to know that the cat was quite unlike any ordinary cat--and cats of any kind are bad enough--and certainly he guessed that the cat under control of its master was one, and away from that questionable influence likely to be another, and very much worse, calamity.

The keeper searched that cottage from chimney to doorstep. No cat there. His dog did not, as might have been expected, help him in this search. Indeed, his dog, he now discovered, had vanished--had, in fact, gone out at the back-door and cleared off.

Meanwhile the cat was, for his sins, being horribly pricked by the holly-hedge through which he was sliding. He growled under the punishment. Ordinary domestic cats do not, as a rule, growl in such cases, though they may "swear."

Once through the hedge, the cat dropped into the ditch on the other side, turned to his right, and galloped up it. It ran upwards, skirting a sloping wet field, to a dark, damp, black wood, as woods always are that stand on cold clay and have much evergreen growth. They remind one of a wet, chill rhododendron forest of Tibet.

The cat's gallop was in itself peculiar, loose, long, his head low, his forepaws straight, his hindlegs trailing out behind. So does the tiger gallop across the jungle glade when the beaters rouse him.

There were other things peculiar about Pharaoh also, now that one had him on the move and could see. He was, perhaps, a fraction big for his kind; his coat was yellowish, fading beneath, with "faint pale stripes" well marked on the sides; his tail was long, and oddly slender and "whippy," ringed faintly to the black tip; his fur was short and harsh, quite unlike that of a domestic cat, and the expression of his eyes was one of permanent, unsleeping fierceness.

Once he stopped and stared back, and in the pause which followed one could distinctly hear a faint but rapidly increasing drumming sound following his trail up the ditch. And least of all beasts had that cat delusions. He turned and galloped on. The keeper's dog was of an independent turn of mind. He had quietly run that cat's trail, forgetting that, in the long-run, dogs are not fitted to maneuver independently, and may suffer if they do so. You see him flying up the trail, square nose to ground, tracking really very cleverly indeed, and with a fine amount of what huntsmen call "drive."

Ho had overtaken Pharaoh before the hunted one could reach the wood. He realized it as he took the last bend in the ditch, when he saw a yellow streak rise under his nose, and bound, with all four legs stuck out quite straight, and claws spread abroad, like a rubber ball out of his path, avoiding his clumsy, murderous snap by an inch, and then felt it rebound right on to his back.

The next few seconds were quite crowded, and that dog had the time of his life.

Even an ordinary domestic "puss" can make wonderful havoc of a dog's back when once it gets there; and stays, as it does, like a burr, and this one could go a bit better than most; and when that dog at last got the cat's "leave to go," he went rather sooner than at once, proclaiming his misery aloud to all the world, so that his master, coming at that moment out of the back-door of the cottage, heard him afar off, and swore.

As for the cat, he turned about, all bristling, and went too. He went straight up to, and through, the wood, disturbing in clouds the starlings, who had just come in to roost in the rhododendrons, so that they rose with a rushing of wings like the voice of a thunder-shower on forest leaves, and incidentally drenched the cat with a deluge of raindrops collected in the leaves as he raced through underneath. A lesser beast, it may be noted, would have climbed a tree, but Hawkley, I think, had convinced his cat of that folly when a man might be following up behind.

Straight through the wood galloped Pharaoh, and into a stretch of age-old furze, or gorse, if you like, beyond. That showed strategy. The furze was a maze of a million spikes, and branches, and twisted, gnarled stems tough as wire-rope; a wonderful place, all honeycombed with rabbit-runs; a world unto itself.

The cat moved on quickly into the heart of the furze, pausing every few strides to listen and glare round. Several times he sniffed the sickly grass and the carpet of dead spikes.

Once or twice something moved ahead; a branch was shaking as he came up, a blade of grass slowly righting itself, as if something that had been sitting upon it had but just stolen away. All round were hints of life, but no life was visible. It was as if the cat were moving through an army of ghosts.

Then, in a flash, without any kind of hint or warning to prepare one for the unnerving contrast of the change, was war--raw, red war.

There had come up a rabbit-run--a regular rabbit-turnpike--a creature. It was strikingly colored, that creature, and big--nearly three feet long, to be exact; but it looked much bigger in the ghostly twilight--and yet till it was actually upon him he, even he, had failed to see it.

Long, low, bear-like, and burly, with claws caked with earth, gashed and bleeding on flank and shoulder it was, red-fanged and wild-eyed. It charged home upon Pharaoh without a second's pause, and with an obscene chatter that was unnerving to any one, let alone so highly strung a bag of tricks as a cat.

Men and dogs had been besieging this badger in its den for twelve hours. It had in the end made a desperate *sortie*, upset one man who had failed to grab its tail, run into and bitten another, and got clean away. Pharaoh was unfortunate in that he stood between the half-mad beast and another den for which it was making.

There was no time to go back, no room to execute one of those beautiful lightning side-leaps which are the pride of all the cats, and less to spring into the air, a neat trick of the tribe which it has also perfected.

The cat was cornered, and, being cornered, fought like--a cornered cat! That is to say, an electrified devil.

He reared up. He struck, pat! pat! right and left, with the terrible, rending, full stroke of his kind. He met open jaws with open jaws--you could hear fang clash against fang. He grabbed, scrunched, drew back, grabbed, scrunched again, as a lion will--for the cats neither hold fast like a weasel nor snap like a wolf. Then, as the full force of the charge and the weight of the enemy's body--some twenty-seven and a half pounds--took him, he hugged, round-arm fashion, with his talons, and, still grabbing and scrunching, rolled over backwards.

Cat and badger turned into a ball--a parti-colored ball, very lively as to its center, and it whirled. Unfortunately there was not much room to whirl in. That made things all the more grisly. You could almost see the grim skeleton shape of death, hovering over that growling, snarling, spitting, worrying, tearing, kicking, gnashing, scrunching, foaming, blood-flecked Catherine-wheel--almost see death, I say, bending down with upraised arm ready to strike. But death never struck.

In an instant there came, sounding strangely hollow in that still, damp air of dusk, as though it were in a cave, the unmistakable noise of a deep, dry, hacking cough. Truly, it was nothing much--just a good old churchy and human cough. But it might have been a blast from the trumpet of the archangel Gabriel himself by the effect it had upon the two combatants. They shot apart like released electrified dust-atoms, and--pff!--they were gone--wiped out. Like pricked bubbles, they had ceased to be. And neither gave any explanation. Being wild things, of course they wouldn't.

The cough had only come from a laborer, who, passing along a pathway through the furze, had heard the commotion, and stopped. He never saw anything, though he crashed into the furze and hunted--he never saw anything, which was no wonder, seeing that he could hardly have selected a way to see less. The cat was four hundred yards away by that time, and goodness knows where the badger was---deep down in his den, one presumes.

Later the cat slept, in a fortress of nature safe enough, surrounded by a hundred unseen sentries with brown jackets and white tails--rabbits, who would give him all the warning he required.

II

The lean night wind next evening came down, and day went out almost imperceptibly. Blackness grew under the furze caverns, and the last glimpse of the estuary faded away in a steely glimmer; a brown ghost of an owl slid low over the spiked ramparts, and wings--the wings of fighting wild-duck coming up from the sea to feed--"spoke" like swords through the star-spangled blue-black canopy of heaven.

The night-folk began to move abroad. You could hear them pass--now a faint rustle here, now a surreptitious "pad-pad" there. Once some bird-thing of the night cried out suddenly, very far away in the sky, "Keck! keck!" and was gone.

It was not Pharaoh, however, that you would have heard move. None of the wild-folk could tell how at midnight he managed to land himself far out over the marsh, unperceived. He was there--you must take my word for it--just two faintly luminous yellow-green lamps floating on the mist.

Not many men knew their way across the marsh by day; certainly not five even of the oldest wildfowlers could have got over safely by night. It was not man, therefore, that was causing the cat to melt into the short, salt grass, so closely that there was nothing of him left. Something else was coming his way.

Along the edge of the dike it came--tall, thin, pale, ghostly, and--yes, I could have sworn it, though night does play odd tricks with the human eyesight--faintly phosphorescent. At least, it seemed to glow ever so dimly, like one that moves in a nearly burnt-out halo.

Every yard or two it paused, that thing. Once there was a splash, as if some one were spearing fish and had missed.

The cat moved rather less than an average stone. He knew that in the wild to be motionless is, in nine cases out of ten, to be invisible. The tenth case doesn't matter, because the creature that discovers it usually dies.

Moreover, there was no cover to move to, and cover is the cat's trump card.

Now, everything would have gone off all right if--well, if the cat hadn't been a cat, I suppose; that is, if he had been able to stop the ceaseless twitching of the black tip of his tail. Tiger-hunters know that twitching, and those who have stalked the lion will tell you of it, as also the sparrow on the garden wall, whose life may have been saved from somebody's pet "tabby" by that same twitching. It is a characteristic habit of the tribe, I take it.

The luminous ghost-thing was close now. Heaven knows whether it saw that twitching then! I think so. It stopped, anyway, and became a pillar of stone. The cat, almost under it, fairly pressed himself into the grass.

Then--whrrp!

Something shot through the air like a lance, and pinned that twitching tail-point to the ground. There had been no warning--nothing! Just that javelin from the ghost, and---the cat on his hindlegs, screaming like a stricken devil, clawing at the ghost, now revealed as a very big, long-legged bird which flapped. It flapped huge wings and danced a grotesque dance, and it smelt abominably, with the stench of ten fish-markets on a hot day.

Then at last, the cat clawing and yelling the whole time, the bird's slow brain seemed to realize the mistake. The javelin, which was its beak, was withdrawn from the protesting tail-tip hurriedly--to be driven through the cat's skull as a sheer act of necessary self-defense, I fancy. But the cat did not wait to see. Imagine the infamy, the absolute sacrilege, from a cat's point of view, of spitting a feline tail in that disgusting fashion. Why, if you only tread on one, you hear about it in five-tenths of the average second, and offend the supercilious owner for a month afterwards!

There was a vision, just a half-guessed vision, of our cat shooting straight upwards through the air, and outwards over the still waters of the dike; there was a number one splash that set the reflected stars dancing, and the water-voles ("rats," if you like) bolting to their holes; and there was the sighing "frou-frou-frou!" of great wings as the big bird rose and fled majestically. There was the sucking gurgle and drip-drip of a furred body leaving the water on the far side, eyes that glared more hate than pen can set down, and a deep, low, malignant feline curse. That cat had swum the rest of the way over the dike which he could not jump.

The bird was only a heron, and that does not sound much unless you are acquainted with the ways of the heron and all his beak implies. A heron is one of those birds that can fight at need, and--knows it. Moreover, in his long beak, set on his steel-spring neck, he has a weapon of awful "piercefulness," and--knows that too. The bird is an example of armed defense.

This one had merely been fishing for eels in that pessimistic way peculiar to all fishermen, and seeing the tail-tip waving in the grass, and nothing else, had mistaken the same for his quarry. And this will be the easier to believe because we know, and probably the heron did also, that eels are given at times to overland journeys on secret errands of their own.

The cat crawled away down the dike in offended silence. He was wet, and the only cat I ever knew who did not seem to be scandalized past speech at the fact. Indeed, he went farther. He came upon a ripple and a dot, some fifty yards farther on, which to the initiated such as he, represented a water-vole ("rat," if you will) swimming.

Then, before you could take your pipe from your mouth to exclaim, the water-vole was not swimming. He was squealing in a most loud and public-spirited fashion from between Pharaoh's jaws, and it was the cat who was swimming. He had just taken a flying leap from the bank and landed full upon the dumbfounded water-vole--splash! Then he swam calmly ashore and dined, all wet and cold. Now, what is one to say of such a cat?

"Landed full upon the dumbfounded water-vole—splash!"—*Page 90*

[Illustration: "Landed full upon the dumbfounded water-vole--splash!"]

III

Long did the keepers, in Colonel Lymington's woods and along the hedges, search with dog and gun for Pharaoh, and many traps did they set. The dogs truly found a cat--two cats, and the guns stopped them, but one had a nice blue ribbon round his neck, and the other had kittens; the traps were found by one cat--and that was the pet of the colonel's lady--one stoat, one black "Pom"--and that was the idol of the parson's daughter--and one vixen--and she was buried secretly and at night--but Pharaoh remained where he had chosen to remain, and he remained also an enigma.

Then the colonel's rare birds began to evaporate in real earnest. Hawkley's little efforts at depleting them were child's-play to those of Pharaoh that followed, although, of course, Pharaoh himself did not know, or care the twitch of a whisker, whether the birds he slew were rare or not.

Now, if there was one thing more than another about which the colonel prided himself in his bird sanctuary, it was the presence of the bittern. I don't know where the bittern came from, nor does the colonel. Perhaps the head-keeper knew. Bitterns migrate sometimes, but--well, that keeper was no fool, and knew his master's soft spot.

It was a night or two later that Pharaoh surmounted the limit, so to speak, and "sprung all mines in quick succession." He had been curled up all day in his furze fortress, that vast stretch of prickly impenetrability which, even if a dog had been found with pluck enough to push through to its heart, would still, in its massed and tangled boughs, have given a cat with Pharaoh's fighting prowess full chance to defy any dog.

He was beautifully oblivious of the stir his previous doings had kicked up, and of the winged words the colonel had used to the head-keeper; of the traps set all about, of the gins doubled and trebled in the wood and round the park, and of the under-keepers who, with guns and tempting baits, took up their positions to wait for him as night fell.

No one seemed to have suspected the furze a mile away, and still less the marsh and the coverless bleak shore of the estuary, as his home. Indeed, no one looks for a cat on a wind-whipped marsh when woods are near at all. Yet this open, wet country seemed to be a peculiarly favorite hunting-ground of Pharaoh's.

It was a night of rain-squalls and moonlit streaks when Pharaoh, wandering devious among the reeds, first became aware of the bittern, in the shape of reptilian green eyes steadily regarding him from the piebald shadows. Possibly the cat's whiskers first hinted at some new presence by reason of the "ancient and fish-like smell" which pervaded this precise reed jungle.

Pharaoh stopped dead. Pharaoh, with cruel, thin ears pressed back, sank like a wraith into the soft ground. Pharaoh ceased to be even a grayish-yellow, smoky something, and became nothing but eyes--eyes floating and wicked. A domestic cat, after one frozen interval, would have crept away from the foe it could not fathom, but Pharaoh had other blood in his veins.

To begin with, he was wondering what manner of beast the owner of those saurian-like orbs might be. To go on with, he was hungry, and--smelt fish. But though he was looking full at the big bird, he could not see it, which is the bittern's own private little bit of magic.

Nature has given him a coat just like a bunch of dried reeds and the shadows between, and he does the rest by standing with his bead stuck straight up and as still as a brass idol. Result--invisibility.

None know how long those two sought to "outfreeze" the "freezer," while the rain-showers came up and passed hissing, and the moon played at hide-and-seek. None knew when Pharaoh, flat as a snake, first began that deadly, silent circling, which was but acting in miniature the ways of the lion. None knew, either, at what point of bittern first begun to sink and sink, till he crouched, and puffed, his neck curved on his back like a spring ready set, his beak, like a sharpened assegai, upright.

Only the short-eared owl, with his wonderful eyes, beheld Pharaoh make his final rush; watched that living spring *sprung* quick as light, shooting out straight at the cat's glaring eyes, and saw--greatest miracle of all the lot--Pharaoh dodge his head aside in the twentieth of a second, and blink, letting the blow that spelt death whiz by.

And only those same owl's ears--sharpest of all the ears of the wild--heard the diabolical yell of Pharaoh as the long, sharp beak pierced through the loose skin of his shoulders, and, thanks to that same looseness, came out again an inch or two farther on, transfixing him; or listened to the devilish noise of the "worry," as the cat turned in agony on himself and buried his fangs where he could behind those expressionless green reptilian eyes; or caught the stupendous flurry and whirl of wings and fur and gripping claws and scaly legs, as a cloud put out the moon and darkness fell with silence, like the falling curtain that ends a play.

The very last pale rays of a watery setting sun slid bar-like through the cottage window, and fell, twirling, aslant the floor.

A late spider had spun a web across the fireplace, and the one last fly that always lingers sat in the sunbeam. It was Hawkley's cottage, dismantled and derelict.

Something like a furry round hassock, lying motionless in a far corner, moved at the sound of rain, and lifted a round head with round eyes that glared with so terrible an expression that one caught one's breath. There was blood--dried blood--by the furry shape, and drops of dried blood across the floor from the window in the next room, that it had been nobody's business to shut.

The day went out in gloom and howling rain-rushes. Darkness took possession of the room. And--*the gate clicked.*

Truly, it might only have been the wind, but--Pharaoh was on his feet in a flash, growling, and there was a glint of green-yellow light as his eyes whipped round.

Followed a pause. Then, in a lull, once, twice, the unmistakable crunch of a shod foot on gravel.

Another pause. Pharaoh was crouching close now, trembling from head to foot.

"Pharaoh! Pharaoh! Pharaoh, old cat, are--are you in there?"

The voice, strained and husky, came in at the open window. In the last lingering afterglow of dying day, a face, haggard and set, showed there, framed in the lead casement.

"Phar---- Ah!"

Pharaoh was up. Pharaoh had given a strange, coaxing little cry, such as a she-cat gives to her kittens. Pharaoh, lame and stiff, but with tail straight as a poker, was running to the window in the next room, was up on the sill, was rubbing against and caressing the haggard face like a mad thing.

There was a long, tense pause, broken only by a continuous purring. Then the creaking sound as of the lid of a wicker basket being opened. The purring ceased. The creaking came again, as if the lid were being shut. There came the crunch once more of stealthy shod feet on gravel, the click of the gate, and--silence!

Hawkley had come for, and found, his cat.

VI

THE CRIPPLE

It was gradually getting colder and colder as he flew, till at last, in a wonderful, luminous, clear, moonlit sunset, when day passed, lingering almost imperceptibly, into night, the wind fixed in the north, and a hard white frost shone on the glistening roofs--far, far below.

Up there, at the three-thousand or four-thousand feet level, where he was flying, the air was as clear and sparkling as champagne, and as still as the tomb. If he had been passing over the moon instead of over the earth, the effect would have been something like it, perhaps.

He was only a thrush, *Turdus philomelus* the songster, but big and dull and dark for his kind, and he had come from--well, behind him, all shimmering and restless in the moonlight, like a fountain-basin full of quicksilver, lay the North Sea; ahead and beneath lay England; and across that sea, three hundred miles, as I count it, at the very least, to the lands of melting snow, he was going when late cold weather had caught him and warned him to come back. And alone? No, sirs, not quite. Ahead, just visible, blurredly--a little phantom form rose and fell on the magic air; behind, another; on his right, a third--all thrushes, flying steadily westward in silence; and there may have been a few more that could not be seen, or there may not.

His crop, as were the crops of the others, was perfectly empty. Indeed, he appeared to prefer traveling in ballast that way. But his eyes shone, and his wing-strokes, with little pauses of rigidity between, such as many birds take--only one doesn't notice it much--were strong and sure.

Once a large-winged, smudged shape, making no sound as it slipped across the heavens, came flapping almost up to him, peering this way and that at him and his companions, with amber flaming eyes set in a cat-like, oval face. The thrush's heart gave a great jump, and seemed threatening to choke him, for that shape--and it howled at him suddenly, in a voice calculated to make strong men jump--was death of the night, otherwise a short-eared owl.

But a gun went "boomp!" with that thick, damp sort of sound that smacks of black powder, somewhere down on earth, and a huge "herd" of green plover, *alias* peewits, which are lapwings, rose, as if blown up by an explosion, to meet them, their thousand wings flickering in the frost-haze like a shower of confetti, and the owl was so disconcerted by the disturbance that he dropped back into the night whence he came, as one who falls into a sea.

Then suddenly the thrush--all the thrushes, indeed--tipped tails, and flew downward--offering no explanation to help one to understand why--till they dropped, each one entirely on his own hook, apparently, in or about some gardens, as if they had tumbled out of the sky; and our thrush, in twenty seconds, had slipped into some apple-trees, and thence to some laurels round a shed, and--was asleep! I say "was asleep." Out of the starry sky, down, in under, and asleep--all without emotion, and like a machine. Now, what is one to make of such a bird?

He did not see, or, more correctly, did not appear to see--for I do not know what he saw and what he pretended not to see, really--the lean, lithe, long, low weasel that passed, climbing and sniffing, beneath him--within six inches--possibly scenting out a rat. He did not hear, or show that he heard, the blackbird--she was rusty, dark brown, as a matter of actual fact--scream, a piercing and public-spirited scream, when the very big claws of a little, round, spotted-feathered ball with wings, like a parody of a cherub--but men call it a little owl, really--closed upon her and squeezed, or pierced, out her life. He did not feel, or let on that he felt, the branches gently sway as two eyes, glinting back the light of the moon--eyes which were the property of a "silver tabby" female cat--floated among the twigs, looking for him, him most certainly, whom she seemed faintly to smell, but never saw.

"A 'silver tabby' floated among the twigs, looking for him"
—Page 98

[Illustration: "A 'silver tabby' floated among the twigs, looking for him"]

These things represented tense moments dotted through hours of cold, dark silence, and the blue-black dome of night arched, and the moon drifted, all in rigid, cold, and appalling stillness.

Then the wind changed, and our thrush awoke to a "muggy" day, under a soaked, cotton-wool, gray sky, all sodden with streaming showers of rain. And, by that token alone, he must have known that he was in England. No other climate is capable of such crazy, unwarned, health-trying changes. He had come in an icy, practically petrified silence. He left in a steaming, swishing, streaming gale.

But that was not before he had been down to scratch like a fowl among the dead leaves under the privet-hedge for grubs, who "kidded" themselves that they were going to be fine, flashing insects next summer. He also prospected a snail or two, and broke through their fortifications by hammering the same upon a stone. And, by some magic process that looked akin to the way in which some men divine water, he divined a worm out of seemingly bare earth. It was there, too, and it came up, not joyfully, but tugged, to be hammered and shaken into something not too disgustingly alive to be swallowed.

Then, while a robin mounted to a spruce-spire and acted as Job's comforter to all the birds of the garden by singing--ah, so plaintively and sweetly!--of the dismal days of frost and snow, he "preened"--i.e. went over and combed every feather, and tested and retested, cleaned and recleaned, each vital quill. Then, in one single, watery, weak stab of apology for sunshine, on the top of a fowl-shed, he surrendered himself to what, in wild-bird land, is known as the "sunning reaction," which really consists of giving body and mind utterly to the sun and complete rest.

And then he left.

Now, it was no chance that he left. Birds don't do business that way. To you or me, that location and its climate would have seemed as good for him to "peg out a claim" in as any other. He knew better. Something--Heaven alone knows what--within him told him what was coming. He had the power to take a draft on the future, and by that means to save himself--if he could. Wherefore he flew on southward--always south.

And six hours after he had gone, the wind swung like a weather-cock, swung and stopped at northeast, and frost began to grip that garden in an iron fist that threatened to squeeze the life out of every living thing in it, and the sky hung like the lid of a lead box.

The thrush flew, with a few halts, practically all day and well into the night, and the northeast wind and the Frost King chased him south.

He roosted in a great fastness of age-old holly-bushes within a wood, whose branches were packed with his relations--redwings, thrushes, and blackbirds, and also starlings--all tired out, all booked for the south.

Some woods seem to hold a curse of gloom. One cannot say why. And this was one of them. And the tawny owl that nobody saw but everybody heard, and the white stoat that everybody saw and nobody heard, and the amorous dog-fox with the cruel bark that everybody saw and heard, did not, taken together or singly, add to the gayety of the scene.

The thrush was just ahead of the cold when he went to roost in pouring rain. In the night, however, the cold had overtaken him, and the thousand-jeweled beauty of frost-flakes flashed to his waking eye.

He was numbed and puffed out and peevish, and disinclined to move, but anything was better than sticking about in this roosting-place, this casual ward and clearing-house of the wild. The keen starlings were already off, swinging away, regiment by regiment, with a fine, bold rush of wings; the blackbirds were dotting the glades; the redwings were slipping "weeping" away, to find soft fields to mope in; and the pigeon host--what was left alive of it after diphtheria had taken its toll--had streamed onwards, heading southwest.

Turdus philomelus spelt L-u-c-k for our friend that morn, for he had not prospected two hundred yards when he came on a place where a vagrant "sounder" of half-grown, domestic, unringed pigs had been canvassing the wood for beech-mast, acorns, and roots during the night. The soil was all torn up for a space of about an acre, probably the only soil for miles--except along streams and by springs--penetrable by beaks until the sun came out; and the thrush feasted royally upon hibernating caterpillars and chrysalids that would have become moths, beetle larvae all curled up and asleep, and other pests; and he must have done a considerable amount of good in that place during the next hour or so.

But feasts do not go begging long in a frost-bound wild, even if they are hidden; and by the time our thrush had driven several other thrushes away--for he was a jealous feeder--and had been driven away by blackbirds himself more than once, starlings descended upon the place with their furious greed, and our thrush concluded that it was about time to "step off." The crowded place might become a quick-lunch resort for some others, not insect-feeders--hawks, for instance--and was unhealthful for that reason. Indeed, he had not more than moved away into the shelter of the rhododendrons when a shadow with a hooked bill shot round the corner, going like the wind. He had time to see it dive like a dipping kite--but it was a sparrow-hawk--and to hear the death-scream of a feeding blackbird, before he went completely from that place, and it knew him no more.

Soon after that he sighted the sea, wide-stretched and restless, ahead, and turned westward parallel with the coast-line, till, in the afternoon, he came unto "a land where it was always afternoon"--a flat, damp, dwarf-treed, relaxing, gray land, mild, as a rule, and melancholy--a land full of water. But for once it was a cold land, and the thrush realized that the bitter frost had leapt ahead of him, and that he might now never outstrip it again, perhaps. I do not know if he realized, too, that the lead sky, that looked as if it were going to come down and crush one, meant snow.

In a bare orchard he was attracted by the sight of several blue titmice and two robins, feeding upon one or two odd apples that had been left unpicked at the very top of a tree. It seemed strange and out of place to behold apples in midwinter like that; but, for some reason, he took only a few pecks, and his devil prompted him down to peck at some soaked bread among the violets, and to drink at a spring so exquisitely encrusted with moss that it looked as if everything, every floating dead leaf, stone, and root, had been upholstered in plush.

Then Fate struck--hard.

A snap, a thump, and he was bouncing over and over, with an air-rifle bullet in his thigh. It was a blow that knocked him half-silly, and he was down before he knew, but only for a second, because of what he saw. He beheld a boy, with an air-rifle in hand, running towards him; but ahead of the boy was the boy's young cat, who evidently had learnt to look for a meal when the air-rifle went off.

The cat, being young, however, managed to bungle his pounce for the fraction of a second, and that is long enough for most of the wild-folk. Came a mad fluttering, a beating of wings, a quick mix-up, and, before he knew, that cat found himself frantically chasing that thrush across the orchard, striking wildly always at a thrush that just wasn't there, as the latter part flew, part hopped, with every ounce of strength and agility that clean, hard living had given him, till he was clear of the trees. Then--up and away, with his heart in his beak, so to speak, and his brain whirling, till the orchard lay "hull down" on the horizon, and was only another bitter experience, and a warning, seared into the bird's memory.

So far, so good. He had made his escape, had euchred Fate, but--the payment for laziness, the terrible cess for a momentary lapse from vigilance, which great Nature, in her grim, wise cruelty, always demands, had to be met, and the end of it was not yet.

It began, however, now.

The thrush discovered that he was not alone in the air, and that he had all at once got himself, as it were, fixed in the public eye, and was "wanted." A swish in the sky made him look up, to see a rook, with a leering eye, coming down upon him. He cleverly "side-slipped" in mid-air, and let the rook, braking wildly, go diving by. Perhaps he wondered what had turned the rook hawk. As a matter of fact, the weather had, partly, and the rifle had, the rest; for the rook could see what the thrush did not yet realize.

The rook went away astern, shouting bad language, and another foe came to take his, or her, place. Again our thrush discovered that he was not alone. Little, white, silent, cruel, dancing flakes of white were traveling more or less with him and downwards, upon the following wind. The snow! The snow at last! And he was trapped, for it was to keep ahead of the snow that he had journeyed all that way back again. Indeed, you can hardly realize, unless you have almost lived their life, what the snow and the frost mean to all the thrush people, but more especially to the common song-thrush and the redwing. At the worst it means death; at the best, little more than a living death.

However, to race the snow were useless. Yet he flew on, and on, and on, like a stampeded horse, blindly, one-sidedly, while the ordnance survey map beneath turned from brown, and chocolate, and silver-gray, and dull green, first to pepper and salt, then to freckled white, then all over to the spotless white eider-down quilt of the winter returned, as far as the eye--even his binocular orbs--could reach, muffling tree and house, and garden and copse, and farm and field, and fallow and plow and meadow in the one mystical, silent, white disguise of winter. And the thrush at length came down.

His eye had spotted a little corner of a garden that might have been a spread table in the wilderness. It was only a small triangle of lawn, with a summer-house at its apex, and a spruce-fir and a house at its base, and privet-hedges marking off the rest. But it had a "bird-table," and a swept-clean circle on the grass, and there was sopped bread upon both. And that place was given over entirely to chaffinches, *all hens,* tripping, mincing, pecking, feasting, fighting--because they were chaffinches, I suppose, and must fight--all over the place.

The thrush came to anchor upon the roof of the summer-house, and--straightway fell upon his beak! And that was Fate's punishment for laziness, one second's relaxation from vigilance.

Righting himself, he almost overbalanced the other way, and only finally managed to come to an intricate halt on one leg. The other leg--the right one--was twisted back under him, in line with his closed wings and tail; that is to say, it was pointing the wrong way for a bird's leg, or, rather, so far as could be seen among the feathers, that was how it seemed. But the leg was not broken; he could still move his toes and expand his foot. Otherwise he could do nothing with it. The leg might not have been there, for all the use it was to him; it would have been better if it had not been there, for it hampered his flight, or unbalanced him, or something, so that he was incapable of traveling now beyond the snow, even if he would. Undoubtedly the air-rifle had done its work.

Now, in the wild it is a fairly sound maxim that an injured wildling is a dead wildling--that is, unless the injury is quite slight. There are exceptions, of course. Flesh-wounds and quick-healing wounds are exceptions.

However, our thrush seemed to be no coward, and he at once buckled to, to fight Fate and all the world--one bird *v.* the rest. It was appalling odds, and I guess no darn fool could have been found to back that bird's chance of winning through.

Then he showed that he had at least one trump up his sleeve. A shape like unto the shape of a silken kite came floating in ample circles across the low-hung sky. And the color of that shape was brown--pale brown; and the shape was alive, and had the appearance of eternally looking for something, which it always could not find. So hunts the kestrel falcon, and by the same token the thrush knew that this was a big hen-kestrel. I say "big" advisedly, because in kestrel society it is the ladies who have the weight and the vote.

And the thrush, who had by that time flown to the ground, promptly "froze "--froze to stillness, I mean--and vanished. It was a startling little trick of his, almost an eccentricity; but the fact was that so long as he kept still on the dark ground where the snow had been swept away--and earth and grass mingled almost to a black whole against the white--he was practically invisible. This was because of his peculiar somber color. Had he been light of dress, like an ordinary song-thrush, any eye could have picked him up in that spot.

Now, that kestrel was in a bad temper and vicious. She was cursing the snow which covered the doings of the field-mice, which ordinarily were her "staff of life"; and she had not killed since dawn. Hence she was a public danger, even to wild-folk she usually left alone, and just now she was looking for our thrush, who she had seen fly down and--vanish.

There he was, however, bang in the open, unshielded by any cover, motionless on one leg, looking upwards, and, to all intents and purposes, not there. The kestrel came shooting up superbly, going at a great pace on the wind, cutting the cold air like a knife, twisting and turning her long tail tins way and that, but moving her quarter-shut wings not one stroke. Right over him she dived, her wonderful eyes stabbing down, so close that you could see her small, rounded head turning and craning. But no thrush did she see. She "banked," hung, swept round, and came back. Then she hovered, like a bird hung from the sky by an invisible hair; and for our thrush she was indeed the sword of Damocles, for the spot in the air where she hung was directly over him. If anybody had shot her dead at that instant, she would have fallen upon his back. At that instant, or the next, she might fall upon his back, anyway, without anybody shooting her. Indeed, the betting seemed a good few hundred to one that she would.

Very few human beings know the full meaning of the word "still"--not even bluejackets!--but most of the wild-folk do. They have to. So did the thrush, but never before had he kept so utterly, stonily, frozenly, strickenly motionless. If he had moved an eyelid even, winked, or gulped too hard, it would have been all up with him. But he didn't and it was not all up; though the kestrel seemed as if she were going to hover there, in that spot, through all eternity. And when at last she condescended to surrender to the wind and vanish like a falling star into the horizon, our friend was as near nervous prostration and hysteria as a bird can be. A very little longer and I believe he would actually have died from sheer overstrain, instead of from kestrel.

Then the thrush fed. He did it against time, before dark, for if night came and caught him with an empty crop, he froze. Perhaps he would freeze, anyway; but no matter.

The hen-chaffinches, presumably at the end of a journey, or part way along it, too, were in a like hurry, and for the same reason. He could see them now only as faint splashes of white, as they opened tail and wing to fight; but they could not fight *him,* and he savagely kept the little clearing in the snow free of all save himself. It was as if he knew that he was "up against it," and the fact had developed a grim fierceness in his character.

An owl must have gone over about this time, because an owl did go over that garden about the same time every night; but perhaps she was not expecting thrushes in that gloom, or was in a hurry to keep an appointment with a rat. Anyway, the owl did not develop.

Thereafter and at last the thrush went to sleep in a spruce-fir.

Dead silence reigned over the garden, and Cold, with a capital C, gripped the land. Heaven help any bird who roosted on an empty stomach on such a night! It would freeze to its perch before morning, most like.

Indeed, our thrush had a neighbor, a hedge-sparrow just newly arrived from "somewhere up north." It had come in after dark, and therefore had no time to feed. The thrush just took his head out from under his wing and opened one eye, as the poor little beggar perched close to him for company. He could see it plainly in the petrified moonlight.

When next he opened one eye and looked, dawn was at hand, and the poor little bird was still there. When at last, with shoulders humped and feathers puffed, our thrush flew down to feed in the first pale-gold glimmer of very-much-diluted sunlight, the hedge-sparrow did not move. Now, in opening his wings, possibly from a vague idea of frightening the hedge-sparrow away from the magic swept circle on the lawn close by, and its bread, the thrush brushed heavily against that hedge-sparrow, so that--oh, horror!--it fell, or swung over backwards, rather, and hung head downwards, swaying slightly, like a toy acrobat on a wire, before it fell, so rigidly and so stiffly immovable that one expected it to shatter to pieces like glass as it hit the ground. It did not, however. But it did not matter. The hedge-sparrow was quite, quite dead before it fell, frozen stiff and stark in the night. And none of the other birds seemed to care. Why should they? Such a fate might overtake themselves.

The thrush, much tucked up, but still with some fight in him, was late. Big flocks of peewits or green plover--he could see them between the spruce-boughs--had gone drifting by, winking like floating silver, high overhead, bound westward; and skylarks were passing over the garden, one by one, heading southwest towards the warm, and chortling to each other as they went. Starlings--some of them with extraordinarily bright-yellow dagger-beaks, and some with dull beaks--were before him, squabbling and sparring over the bread on the lawn. A robin dropped a little chain of melancholy silvery notes, and a great titmouse bugled clearly, "Tingling! Tingling! Tingling!" Some one opened a window of the house giving on to the lawn, and the last house-fly blundered out into the cold air; and a company of gnats--surely the most hardy of insects--was dancing in the pale sunlight by the summer-house, *above the snow.*

The opening of the window had erupted the starlings into the surrounding trees, there to whistle and indulge in a "shiveree," such as is dear to the heart of the excitable, social starling. And our thrush was standing motionless in the middle of the swept circle on the lawn almost at once. No one saw him go there. Indeed, unless the observer looked closely, no one saw him at all, for even then he was, unless he moved, difficult to see, and, whatever had been his custom before, in those days he moved but little.

He had come at even to a garden given over to hen-chaffinches--no cocks, as we said--but at dawn, or, rather, his later hour for rising, he found the garden given over to song-thrushes, all pale beside him, all slim, all snaky of build--Continental song-thrushes, most like, and the same only come to those parts in very hard weather, for they come a long way.

Our song-thrush, standing on his one leg, looked at them with one shrewd eye. There were two of them in the snowless circle on the lawn, which had been swept clear of the snow, that was now deep, before he was up, and had also been replenished with bread. Two thrushes sat in the spruce-fir, and one on the top of the summer-house, and every jack of them was ravenous. He could expect no mercy from *them*. They must live, if they could, and there was not enough food for all. And he asked no mercy himself, either. Still, it was long odds.

Then he showed that he, even a bird, knew the laws of strategy, the essence of which is surprise. He surprised everybody by suddenly charging at the thrush on the lawn near him with a murderous ferocity that took one's breath away. It certainly would have taken away that of the other song-thrush, if our friend had not knocked it out of him by the impact. By all the laws of precedence, of course, any one of those others ought to have sent him, with his one leg, into headlong retreat by merely threatening. But our friend was not concerned with the laws of precedence, it seemed. He became a law unto himself, and a most amazing "character" to boot. Also, he fought like several demons, and, by sheer reckless fury, removed that dumbfounded rival of his from the lawn in twenty-one hectic seconds.

Then he fed--it was enough only to glance, just glance, at the other thrushes and the chaffinches, after that astounding exhibition of his character. He fed, and, after he had stuffed full, he stood still a little way off.

This was the signal for two of the thrushes in the spruce-fir to flap down to the bread. One got there. The other saw what was coming, and turned hastily back. The one that got there snatched up a piece of bread. But he never ate it. Something hit him on the side. It felt like the point of a skewer, but it was our thrush's beak, really, and by the time he had recovered from that blow he found himself so busy saving his eyesight that he was glad enough to drop his bread and go.

That, however, was not enough for our thrush. He appeared to "see red," and with a terrible cruel, relentless "redness." He followed the retreating foe to the spruce-fir, flying heavily and awkwardly by reason of his smashed leg. He perched beside him on the branch he settled upon, nearly overbalancing, and perilously swaying and wobbling, with wings wildly flapping, and he drove that thrush to another branch, with such a rain of pecks that the feathers flew. Nor was even that enough. He followed up the attack, and hustled the thrush from that other branch, so that he flew down the snowed-up road. Then our cripple, spinning in a whirl of snow, hurled himself upon the other thrush in the tree, and drove him out of it into the road.

But even that did not suffice him, for devils seemed to have possessed him, and the thought of opposition sent him crazy. He blundered into the privet-hedge, and unearthed a half-frozen *confrere*, who fled, squawking peevishly, leaving one tail-feather in our friend's beak; and finally he flew down to the road.

In the road, he first of all buried his face in snow, then fell on his side, deep snow not being, he discovered, an ideal medium in which to get about on one leg. During that performance his rivals could have abolished him five times over if they had had the heart to unite. But they seemed to think otherwise, and had not the heart for anything. They sat still, with that helpless abandon that afflicts fowls and other birds in disaster, and they seemed about to starve practically on the spot, if left alone.

Our thrush, however, did not leave them alone. They were a direct threat to his only line of communication with life, so to speak--namely, food. Wherefore, either they or he must go. Soon he found that cart-ruts make convenient roads for the birds in the snow, or perhaps it was the chaffinches, who were following one another in lines along the cart-ruts, who showed him.

Then and there, in the road, our thrush seemed to go berserk. He landed upon the thrush nearest to him, spread-eagled and hammering like a feathered devil. There was a whirl of brown feathers and finely powdered snow for about ten seconds, at the end of which time that other thrush detached himself and fled, oven as his conqueror hurled himself upon the next bird.

There were two here, side by side, but neither was quick enough to parry our friend's lightning lunges, after he had beaten down their guard with his wings; and they, too, got up and winged into the leaden, frowning sky. The others did not wait. They had seen all they wanted to, apparently, and would take no part in the play. They faded out among the drifting snowflakes, over the still, white fields, and our thrush was left to the lawn, and the bread, and the swarming chaffinches, whom he easily kept aloof, and--yes, there was no getting away from it--the one thrush on the summer-house who, you will note, had never moved. But when he looked he found that thrush was not on the summer-house, but on the lawn, eating bread; and when he flew down to the lawn to investigate--he flew and landed very clumsily--he made a discovery that seemed to surprise him; or did he already know it? Anyway, the thrush on the lawn was a lady, and--well, what would you? The cripple balanced as well as he could, and looked foolish. It was all he could do.

The day passed swiftly, and faded out in blinding snow. Most of the time the cripple stood motionless, watching his companion and guarding his swept circle, and, as often as he could, he fed. And neither then nor at any other time, except once when the gardener nearly trod upon him before he would move, did he utter a sound. The last glimmer of day showed him still at his post, motionless, all but invisible. But he roosted, as a matter of fact, in the privet-hedge, on the south side of the summer-house, and this time he was not alone.

The day had been trying enough, with its fights and its three cats, which passed within reach of him, and could have slain him--for his injuries made him slow to get under way--if they had not failed to see him, because so still. The night, however, was a clouded terror.

Certainly he went to bed--if one may so call it--full, if not warm exactly; but that was the only advantage. It snowed with ghastly, relentless steadiness, and it blew like the hacking of sharp knives.

But through it all, because full fed, the cripple, with all his handicap, and his lady companion lived; lived to see the hard dawn pale tardily; lived to watch the kind gardener--under strict orders assuredly, or he would never have done it--sweep a space clear on the lawn and spread food for the birds; lived to ruffle his feathers and fly down; and lived to see the thaw which came that afternoon, when the warm sou'-wester came romping over the land, and winter's last stand was overcome by the forces of spring, and all the wild breathed a sigh of relief and went abroad gayly to feed.

But the cripple lived to see other things. For there came a day, about a week later, when our cripple, who had been "keeping company" all the time with his lady friend, heard the whole dawn awaken to a sudden mighty chorus of thrush song. I don't know why they all chose to burst into song thus as at a given signal, but they did, and the effect upon the cripple and his companion was curious. He had just landed upon the top of the summer-house on his one leg, in a particularly awkward and unbalanced manner, and he perched, listening, as if rooted to the spot, and with something nearly approaching horror in his eyes, it seemed to me.

The female bird listened, too, for about a minute, and then, ignoring the poor cripple as if he had never existed, hopped towards the spruce-fir--atop of which a particularly fine and strong-voiced songster was warbling--as if she were drawn by ropes. And--oh, horror!--the songster came down to *her*.

The cripple never uttered a sound, not a song, or a call, or a sign. He hurled himself straight at this new rival like a bolt shot from a crossbow, and he fought. My word, how he fought! But this new antagonist was no half-frozen, half-starved Continental song thrush. He was a Britisher, thick-set, bullet-headed, thick-necked, who had wintered, perhaps, in the south of Ireland or farther, and he fought like a Trojan.

All up and down the lawn the fight raged, and in and out of the hedges, into the mountain ash and out again, down to the ground and up again; but in the end--ah, but it could have only one end!--the Britisher was on the top of the summer-house, literally shouting his song of triumph. And the cripple was on the ground at the foot of the hedge, beneath the spruce-fir, lying on his side, blood-stained and panting. Nobody saw him creep away. Nobody cared--certainly not his lady acquaintance, who was too busy receiving glad eyes from the conqueror.

Also, nobody saw him die. Yet next morning he was dead, stiff and still on the ground beside the summer-house. Some think that it was the injuries he received in his last great fight that killed him. I do not. I could find no wounds upon him sufficiently severe to sustain that theory. I think he died of a broken heart. Don't you?

VII

"SET A THIEF"----

Cob arrived in a snowstorm of unparalleled ferocity. He came upon extended vans sixty-nine inches from tip to tip, which he seemed as if he were never going to flap. All black above, all white below, he was. The fact was worth noting, because, as seen from below, he looked neither black nor any other hue, but just indiscriminate dark, unless he swerved against the little light, and then his white "hull" shone like silver.

In his calm tacking, in his effortless play, in his superb mastery of the furious gale, one realized that here was one of Nature's masterpieces. He arrested the gaze with his serenity, and in his majesty of flight marked himself as a bird apart.

Here was a bird accustomed to power, to respect, and to wield fear, as a king might do; but he was no king, even among birds. He was a great black-backed gull, immense, austere, and cruel, with eyes as cold as the waves whose glitter they reflected, and a heart as implacable as the storm that cherished it; sea-rover, pillager, pirate, swashbuckler, son of the storm in whose fierce buffetings he rejoiced, master of the gale upon whose fury he flourished--the very spirit of the ocean's frontiers, arrayed in the spotless uniform of the sea, sailing under her bold colors.

And then, as he suddenly came, the watcher, had there been one, would have looked at him expectantly, for an eagle, bristling with weapons, so to speak, fierce-eyed, mighty, and scowling, came flapping heavily across the white-fretted bay. There is expression in birds, and most have their feelings and their character stamped upon their whole body. But there was no expression in Cob. His cold eyes continued to stare with steady stoniness, his vast vans to waft an occasional shallow, lazy quarter-flap, his spotless head to peer down at times. Once only, as the real king of the birds, on his course, drew very near, so that you could hear the deep, dry "hough! hough!" of the powerful wings, did Cob open his red-stained--as it were blood--yellow beak, and give utterance--one could call it no more--and so instantly close his beak again and revert to his absolute expressionlessness that one had a job to realize what, or who, in all that vast scene, had spoken.

"I'm-Great-Black-Back!" he said very quietly, quickly, gratingly, and tersely; and then, as if expecting an answer, added, "Eh?" in a hollow undertone.

The eagle's imperial head jerked round as he flew, and he shot a stabbing, sheathed glance at the great sea-bird, as a king might at a man in a crowd who begins to fumble at his hip-pocket. But, save for that, he took no further notice, and beat on with his terrific, piston-like, regular wing-beats; and the gull, that speckless, dazzling, hardened, hard giant, laughed--laughed, I say, softly and to himself, hoarsely and insolently: "How-how-how-how!" It was as if he laughed in derision.

And then a strange thing happened. From the opposite stupendous cliffs, draped in snow, bejeweled with icicles, frowning and desolate, an ominous black shape flung itself furiously, and made straight for the eagle, barking hoarsely with rage as it came. Another hollow bark followed, and a second evil ebony form hurled down from the tottering cliff-top, and flapped towards the eagle in the path of the first. Bark echoed bark above the deep mutter of the breakers, and the echoes along the cliffs answered both uncannily and mockingly.

They were a raven, disturbed from her wool-quilted nest, and her mate; but if they had been hobgoblins straight from an evil dream, they could not, in that immense, grim setting, have been much more impressive.

The great black-backed gull said no more, but wheeled on as if nothing had happened.

The eagle said nothing, and tried to beat on as if nothing had happened, too. He did not succeed, for the ravens who had been addressing him most particularly soon addressed themselves personally to him; and before he knew just how it all came about, they had summoned a quite amazing and unexpected aerial acrobatic power, and were shooting and diving, striking and flapping, about his regal head in a manner that even *he* could not pretend any longer to ignore. No one, not even a king of all the birds, feels comfortable under the imminent possibility of losing an eye--and such a haughty, wonderful eye, too. Nor did the eagle. And he showed it. One presumes he might have abolished the pair--one or both--but the eagle never let on what he presumed. What he *knew* was that he had nothing to gain in a fight with such super-hooligans, and everything to lose, for one wound only might mean a dead eagle *via* starvation and a dead raven--what was a dead raven worth, anyway, to him, or anybody else?

Therefore the eagle changed his mind about continuing his course, which would have taken him above the ravens' nest. He did it grandly, and without giving the impression that the ravens had anything to do with it--he could have squeezed the life out of them with one awful handshake, if his heart had been as big as his claws. But they had something to do with it. And they knew it. So did Cob, who laughed again, hoarsely and as one appreciating a joke, while he wheeled and wheeled over the following waves, seeing all things and never appearing to see anything.

Then at last, when the king of all the birds had sunk, like a speck of floating burnt paper, away over the far, white-mantled hills, the ravens suddenly evaporated into nowhere. Probably no one had seen them go except Cob, and Cob was by now a lonely, dwindling speck away over the restless ocean. Then he was not. He was coming back, swinging along with great, easy, shallow half-flaps, so sublimely lazy that he seemed merely to swim through the gale. But he covered distance; there was speed as well as majesty in his flight, for all that.

In a very short time he was above the cliffs, silent, sinister, almost stealthy. One of the ravens came back suddenly, diving over the crest, half-demented with anxiety to cover her eggs from that stony stare of the sea-rover; and Cob, seeing where she had come from, surrendered himself to the gale, hurtled down-wind, veered, tacked, circled, rocking, and came down in a series of his oblique plunges--smack-bang into the middle of a gory dinner-party, consisting of the male raven, five gray or hooded crows, and one silver herring-gull, feasting upon the carcass of a dead sheep.

Every head went up, every eye blinked, every wing half-opened, every beak shut tight as Cob, whom everybody had thought to be miles away by that time, threw forward his wings, umbrella-fashion, flung them up, hat-fashion, fanning wide his tail, dropped his giant webbed feet, and came to anchor with a rush. Then he folded those wonderful pinions of his, foot by rustling foot, stared stonily at the amazed, mute company around him, and, throwing back his immaculate, smooth, low-browed, spotless head, laughed to the winds, hoarsely, loudly, wildly--a rude, baleful, transport of mirth:

"How-how-how-how-how!"

The raven did not laugh. He had to feed his sitting wife--not counting his big self--in that bitter weather, and he was pluming himself upon having turned the eagle from sight of this gift banquet from Providence as well as his nest. The gray crows saw no cause for merriment, remembering how big the great gull was, and how small are these little, long-wooled, black-faced hill sheep. Moreover, sheep do not often oblige by getting turned turtle in a cleft of rock, and being unable to right themselves before poor, starving, wild hunters--I won't swear who, but it was not the raven *this* time--can come and peck their eyes out.

Cob looked at them again--all five of the gray crows sitting staring straight down their own black gouge-beaks, hunched, cold, out-at-heels, and dejected. Then he laughed again--burst into another wild, jeering fit of merriment, and fell to work.

First of all, he pointed out to the raven--his beak was the pointer--that he was sitting upon the choicest portion of that sheep, and must make way therefrom *instantly*. Next, he turned his head and looked--only looked--at a gray crow that had presumed, upon the turning of his broad, black back, to recommence feeding, and that hooded crow moved one yard in one second--out of reach. And next, Cob, who apparently loved discipline and cherished good manners, started his banquet, and allowed the others to start theirs.

But it was an unholy feast. Cob tugged and tore like a butcher without any knives. At times he nearly fell backwards, when the meat gave way; at times he bolted, and gulped, and choked horribly; at times he was nearly standing upon his head, and at other times upon his tail; and, in case the others should find the woolly outside, where they alone could feed, too easy, he was continually breaking off, to rush--a red-headed demon from hell now--at the raven, or glare at the crows and remove them yards, as if his eyes could kill. As for the herring-gull, he raced and danced in a crazy circle round his giant clansman, apparently smitten with delirium at the luscious titbits he was obliged to watch vanishing down Cob's bright throat.

The raven, however, was growing desperate. He was under contract to Fate to feed his wife. She would freeze there on her nest in the snow among the icicle-studded ledges else. And every time he had got hold of a big enough dainty to tug free and fly off with, Cob had cut in and collared the said morsel. As a matter of fact, friend raven was a better carver than the sea-pirate, had a beak better suited for the grisly purpose. Finally, the black one got hold of a piece of meat, and did not let go. He hung on, and, before anybody realized that he had moved, Cob's yellow-and-red-painted bill--nearly all red now--had closed upon that raven's neck. There was one wild, asthmatical croak from the raven, a whirl of sturdy black and overshadowing black-and-white wings, and the raven was jerked clean head-over-heels, where, among the heather, he lay for a brief second, kicking ignominiously, on his sable back.

Here the crows fled to strategic positions upon bowlders, waist-deep in heather, hard by, expecting a like fate, and leaving the herring-gull to gobble up what he could in the confusion, and risk his life in the process, when suddenly, above the beating of wings and the hiss of wind, all distinctly heard, and jumped at, the sound of a single, horrible, instantaneous, metallic clash.

Cob's agonized yell, the clash itself, and the whir and rush of wings, as every bird there present literally flung itself into the air, seemed really, though of course they were not, coincident--such is the quickness with which these wild creatures act. But Cob alone remained.

He stopped in mid-spring horribly, and suddenly, as if a Hand had reached up and plucked him back. For a second his wonderful wings beat and beat tremendously, frenziedly, with a noise you could hear all up the hill; then he fell back in one demented, frenzied mix up of bashing, smashing pinions, legs, tail, and whirling feathers.

That clash, which had jarred Cob's frame from head to hind-toe, was a trap, *alias* a gin, *alias* a clam, and the rack of man's Inquisition of the wild. He had stepped upon it; it had gone off, and caught him by the right leg, and, being anchored by a chain, had refused to let him go when he sought to remove himself, trap and all.

What followed during the next minute or two it would scarcely be fair to so fine a bird to print. Moreover, it was unnice to behold. Wild-folk have a habit often of going temporarily insane when they first find themselves trapped, because the trap represents to them the most supreme, the most unbearable, of all terrors--loss of freedom; and freedom is to them more than life, especially to birds, and more especially still to those whose lives are dedicated to the wild, free sea.

At the end of that time Cob lay exhausted upon his side, one mighty pinion pathetically trailing in the snow, his beak open, his whole jet and spotless white body shaken and convulsed with pantings that were almost sobs. He seemed in danger of dying there and then upon the spot, with sheer, sickening horror or a broken heart.

The herring-gull was a silver line--about as big as a thrip--to seaward. The gray crows climbed the heavens to landward, like flies that climb a window-pane. Only the raven had not gone, quite.

The raven was a bird, of course, and every bird has got to do its duty. There can be no shirking. *His* duty was to supply food to keep the fires of life burning in his mate as she sat upon her icy nest. His duty was to see that his eggs, *their* eggs, hatched out; and with him the motto was: "The end justifies the means." This bird, this sea-rover, this big pirate, alone stood between him and the discharge of duty. There was no other way, no other food; he had searched. Wherefore, the raven stayed; he knew all about traps, few better, and he stayed, waiting, if it please you, for Cob to--die!

But Cob would not oblige. He had not got a broken and crushed leg, as the raven possibly expected. He was not injured, as he should have been, according to program; only puffed. The mercy of Allah had seen to it that some teeth of that instrument of vile torture that had hold of him were broken off, and that his leg should have been caught in the gap thus formed. Moreover, the trap had not been looked after; it was rusty, and did not shut quite properly. The spring was weak, or some grit had got in, or something, and a smart rat would have got out of it easily, but a rat is not a gull, and knows too much.

Thereafter, nothing happened for a long while. Cob's first delirium seemed to have spent him, and perhaps taught him how much a leg can hurt when tugged by the full lift of sixty-nine-inch wings, especially when one tries to whirl round upon it when it cannot turn.

The raven sat on his lichen-decorated, snow-draped bowlder, hands in pockets, so to speak, abominably untidy, with a pessimistic hunch of the shoulders, but a light in his eyes, a strangely malignant, devilishly roguish leer, that belied his appearance. Perhaps he was waiting to see if Cob during his struggles obligingly touched off any further deadly surprises that might lie hidden in the vicinity. One never knows. He had seen a gray crow double-catch himself in two traps lying close to one another--once.

Nothing happening, however, that raven presently sailed in on his fine work. He broke his neutrality with a sudden dry rustle of wings, and clumsily half-hopped, half-heavily napped, down to Cob, lying there still and silent, but very much awake, upon the snow. He almost seemed to be rubbing his hands, or, rather, his claws, that ebon rascal. This was, indeed, a game after his own heart.

Cob never moved when the raven arrived. I suppose he knew all about ravens, and what one may expect from them. He only stared at him with one cold eye, a tense, lop-sided stare; and he mouthed a little--if one may be permitted the expression--with his beak, like a man moistening his lips.

The raven looked him over critically, leeringly, insolently, with a hateful air of ownership. Then the raven sharpened the gouge thing which he called his beak--wheep-wheep--upon a stone, as birds do, and tightened his feathers, as if almost visibly tucking up his sleeves for--well, for the job.

Then he tweaked Cob's tail, apparently just to see how much alive he was. But Cob did not move, beyond drawing one webbed leg--the free one--up under him.

Then the raven dug him under the wing--punched him in the ribs, so to speak. But Cob did nothing more than cringe--cringe from head to hind-toe, like a worm.

Then suddenly, startlingly suddenly, with the full stroke, the dreaded pickax blow, of all the ravens, he let drive straight at Cob's clear, shining eye--the left one, with which Cob, with his head twisted, had all along been regarding him. He had disclosed his hand, that raven. It was devil's work.

Till that moment Cob had never moved, as we have said. Save for his one eye and his quivering, one would scarcely have known that he lived. That was his game, perhaps. Who can tell? For a stolid, slow-thinking gull may have, in his way, just as deep, or low, a cunning as a brilliant-brained raven. Anyhow, in that fiftieth of a second allowed, just when it seemed as if nothing could save his eye, Cob's head snicked round and up, and he slid the enemy's beak down off his own with as neat a parry as ever you saw. And he did more. He caught hold of the said raven's beak, got a grip on beak in beak, and once having got hold, he kept hold. This was nothing new to him. It was his way--one of his ways--of fighting rival great black-backed gulls. But it was new to the raven, and he had not previously thought out any proper counter to it. (There is a counter, I think.) Result--caught raven as well as caught gull.

Then it was that raven's turn to go mad, and dance a paralytic kan-kan; but he could not get any change out of that gull. Cob hung on almost as well as the trap hung on to him, and far more twistfully. He was quite at home, of course. He had been brought up to this sort of thing. It was the official regulation gull way of fighting under set rules, but he could rarely get any other bird than a gull to fight with him like it. It was not the raven's way of fighting, though, and I think he felt himself in a trap. He certainly acted like a bird out of its senses, while the gull, flapping hugely, and forgetting, in the excitement, his own bondage, gradually forced the raven's head back and back over his back, till that raven was in the unenviable position of staring over his own back at his own tail, upon which he was ignominiously sitting. Also, his neck was half-dislocated, and he was nearly choking. And about this time it began to dawn upon him that it did not pay in the wild to monkey with great black-backed gulls, even trapped ones. He swore, as well as he could, in a gurgling croak. Then----

Clash!

Horrors upon diabolical horrors! Another trap?

The same ghastly thought flashed to both birds' brains at the same moment, and both literally sprang bodily up into the gale in one maddened leap, both forgetting all else in the panic to be gone.

Both stopped at the same instant, with a jerk that nearly unhinged every bone in their bodies. Both yelled with terror at the identical moment.

Both were released--as by the cutting of a string--at the same fraction of time, and both hurtled aloft at the same fear-blinded, rocket-like speed.

But both had not been caught by the same kind of trap.

It was the jerk that had freed Cob from the really quite light hold, as we have already explained, of the jaws of the steel trap.

And it was the jerk that had torn out some of the raven's tail-feathers, and left them in the jaws of the--gray, old, hill fox.

And it was the fox who was standing all alone, watching, with oblique eyes, the two great birds fast dissolving with every desperate, stampeding wing-beat into the hurrying cloud-wrack and the wild seascape--in opposite directions. He had made a good stalk, but had sprung a little short, had brer fox.

Upon a day, weeks later, we find the raven, whose young had left the nest, stolidly soaring over a small, flat island, golden with furze, purple with heather, pale-rose chiffon where it was covered with sea-pinks.

In addition to these, only one other hue, beside green, was there upon that island gem floating on the jade-green sea, and that was a patch of black and white! It flashed to the eye of the raiding rogue-raven, and he altered course towards it, when it turned into a female great black-backed gull, running, literally racing, to her nest, which the raven could now see, with its two big, buff, dark-splashed eggs.

Down flopped the giant gull upon her treasure, and began yelling, "How-how-how-how!" at the top of her voice.

But the island seemed empty of life, and her yelling useless.

Down dropped the raven in front of her.

Down winnowed the hen-raven at the back of her.

And, both together, they approached. And all the time the great black-backed gull continued to yell, "How-how-how-how!"

At last, when he had got close enough, the cock-raven lunged at her, or, rather, underneath her. She parried his stroke, and--the hen-raven lunged. Nothing now, she knew, could save her eggs, unless she rose to fight the cock-raven. The hen-raven then ran in. She only required a second in which to ruin each egg, but she never got it.

Nobody saw the avalanche coming, but everybody heard it arrive. It was of snow-white, and it was of jet-black, and it knocked the cock-raven one way, and sent the hen-raven, picking up her skirts, as it were, and fleeing, the other. And the name of the avalanche was Cob.

I fancy he considered that he bore a grudge against that cock bandit-raven. Perhaps in dreams he could still feel that trap on his leg. Who knows? He certainly used to wake up with outcries, and he equally certainly made that cock-raven shy of that island for evermore.

VIII

THE WHERE IS IT?

No one would have thought of looking for any living beast in the raffle of dried twigs and tamarisk "leaves" between the crawling, snake-like roots of the feathery tamarisks if it had not been for the noise. The noise was unmistakable, as the noise of a fight always is; and the only other living thing near the spot, a tiny, tip-tailed, brown wren--a little ball of feathers, dainty as you please, and all alone there, and out of place down by the terrible, snow-covered, wind-tortured estuary shore--made shift to remove herself, making remarks--wrens can't help saying what they feel--as she flitted.

Then the combatants fell out--literally. Up from the solid earth between the twisted roots they seemed to come, but that proved the art of one of them in concealing his front-door from the curious, and down the bank of the sea-wall, over and over and over, squeaking the most murderous language, and grappling like pocket-devils--tumbled a little jet-black and a little dark-brown beast.

They continued the duel upon the dry gravel below--the finest and the whitest gravel ever you did see--and they would apparently have gone on for goodness knows how long if a gray-white, thin, worn post a couple of yards away had not turned into a heron and stalked an ungainly stalk towards them.

Then they fell apart, and one of them, at any rate--the brown one--ran away in the shape of a water-vole--water-"rat," if you will--the heron making spear-lunges at him with his bill as he ungainlily skipped at the other's tail all the way up the bank. The other fighter, the black one, could not rightly be said to have turned into anything very much--at least, not anything that any one could swear to. It just seemed as if a dark blur whizzed about--more bird-like than beast-like--around the astonished and prancing heron, and then into nowhere. It was like watching a blue-bottle in a tumbler, and very extraordinary. The heron never even professed to follow it or lunge at it. He preferred the water-vole, whose agility was not too fast to see.

At the place where it had come from, the mouth of the hole, it stopped--this beast that could move quicker than eye could follow--stopped so suddenly and completely that its change from almost lightning motion to stony motionlessness in the fraction of a second was nearly as amazing as its first marvelous exhibition. It stopped, I say, and became a--a rat.

To nine hundred and ninety-nine people out of a thousand, the word "rat" conveys only one impression. This rat did not fulfill that impression. In fact, there is more than one kind of rat, and though fate and their fathers' Kismet has cursed them all with a name of shame, they are not all the kind of people that made it so. There is the foreigner, the invader, the common brown rat, who is accursed; there is the old English black rat, whom the accursed one has nearly wiped out into little more than a ghost; and there is the water-rat, who is not a rat, but a vole, and would thank you to remember the fact.

And this rat was a black rat, as black as jet, shark-jawed, star-eyed, elfin-eared, snake-tailed, lean, long-legged, and graceful--a very greyhound among the rats. He was there, in that dancing-floor of the winds by the estuary, because no common or sewer rats were there. They were anathema to him, and they were worse--death in many horrible forms. He had been there all the summer, all the autumn, and all the---- No, by whiskers! he was not going to remain there all the winter. He had his limit, and he hated cold; and here, down by the flat, sodden, mud-choked shore of the estuary, it was so cold that if you didn't jolly well mind what you were up to and keep your tummy always full, you went to sleep, and--never woke up any more.

A little pile of mussel, winkle, and shore-crab shells, and the backbone of what had been a stranded fish, close to the mouth of the hole, showed the rat's account-book to date; but there was a line to be drawn even in this trade. That dawn--if you could call the gray dark of a snowstorm dawn--he, wondrously adventurous, had gone shell-fish collecting, away out upon the freezing wet mud-ooze. He had got three mussels; a muddy face; muddier feet; nearly an eye pecked out by a mighty, great black-backed gull; three chivyings from herring-gulls; one nip from a crab who ought to have been dead; two winkles under big stones that took half-an-hour to shift; one dead pigmy shrew--length two inches--with a hole in its skull, no brains, and a horrible smell; nearly his life removed by the swoop of a kestrel falcon and the javelin-stab of a heron's beak; and twenty minutes' hard cleaning to remove the mud-stains that were not properly off--to his nice liking--yet. And, to add to that, he had no sooner finished than he found that some clumsy fool of a water-rat--vole, I mean--with a mania for mining, had run a shaft into his hole, and brought the whole roof crumbling down upon his scrupulously neat and tidy nest of fine hay and carefully shredded rush--

the only approximately warm corner he possessed in all that biting cold--so that days of labor would not repair the silly damage; and he had had to enter into a free fight with and turn the fool out, nearly losing his life, for the fourth time that short, dark, bitter day, in the process. And now he had to clean himself all over again!

No wonder he was fed-up, and decided to quit. He loved the dank marsh, the brackish channels, the long, lone wind blowing through the tamarisks, and the smell--salt, seaweed, mud, and fish--of it all; but in this--weather, when the cold here, even in shelter, was greater than the cold in any other spot--and the unchecked wind cut like swords of ice--he realized that one must be an eider-duck or an Iceland gull, a northern diver or an Arctic owl, to stand it, and he was none of these. Wherefore, though the dusk had made the dull day only a little more dark as yet, and the pink, luminous frost-haze still hung in the west, he called down his hole to his wife--his one and only wife, but that was not his fault--and quitted.

Ten minutes later you behold our black rat--if you had eyes quick enough, but it was a matter of momentary glimpses, anyway--trekking up a ditch. You have pretty well got to take my word for it, because, though sometimes you saw him for the half of a second, mostly you didn't, and couldn't tell whether there was his wife only, or he only, or both. Really there were both, but our black friend with the embarrassingly, the abnormally, long tail and the genteel head--Mr. *Mus rattus* on Sundays, if you please, and in nowise to be confused with that *canaille,* Citizen *Mus decumanus,* the common brown rat--had not the slightest intention that any one should see him, if he could help it. His wife might be trusted to look out for herself. And for this reason, perhaps, his march was a progress to wonder at.

Did a flock of wild-duck come over from the sea with whistling wings, he did not so much get under the over-hanging grass as *be* there. Did a "gaggle" of wild-geese go by high over, clamoring like hounds, he went out like a blown candle. Did a party of teal--for it was the magic hour of "flight," when all wildfowl shift their quarters to feed, or not to feed--fairly hissing with speed, like masterless bullets, dash over, he--well, he was not before you could realize that he had moved.

Then up and flew round them a shape, and the name of that shape was death. It might have been a gull with hawk-like form. It might have been a hawk in gull-like light-gray uniform. And it might have been an owl with gull-like dress and hawk-like lines. Whatever it was, it was clipper-built, swift, and in fighting trim. As a matter of fact, it was neither gull nor hawk nor owl, but a harrier, a hen-harrier--that's its name, not its sex, for it was a cock--and the same is a half-way house, so to speak, between hawk and owl. Possibly because they are crepuscular, harriers may be thought more rare than they are. This one was "crepusculing," and--the black rat did not like it. They had met before, and Mr. Ratus had gained a lively dislike for this hawk-owl combination, greater than his respect--which itself was not small--for owl or hawk.

Seeing nowhere to go, and nothing to hide in when he got there, our Mr. Ratus shifted from one spot to the other when the harrier made his cat-like pounce--yes, he was something cat-like, too--and had the pleasure of seeing the harrier's uninviting talons grab a clawful of grass, which, by all powers of judgment, ought to have been black rat's fur.

A mere hawk, or even an owl, might have considered this rebuff enough, but not the harrier. *He* wafted himself ten feet aloft on his long narrow vans, and, flapping owl-like, or almost butterfly-like, began to beat, and the beating of the harrying harrier, up and down, is one of the most trying ordeals, for the game beaten for, in the wild.

Mrs. Ratus sat where she was, he presumed, playing the same bluff.

But both were without cover, and the black rat, I fear, devoutly hoped that she would be fool enough to move and give herself away first; whilst she, on her part, was cursing him for many kinds of a fool for starting their "flitting" before it was dark.

While she sat and froze--in both senses--however, the black rat, rigid as a beast cut out of coal, with one bright, shining eye upon the harrier beating up and down, was probing the dusk with the other eye. And presently he thought he saw his chance. He would have had to move, anyway, I fancy, for the strain of sitting there bang in the open was unendurable. His nerves would have snapped. So he went to his chance--a hole in the bank of the ditch.

I say he went, but I only take it to be so because he got there. One could not actually see him go. One had only an idea, quite an uncertain idea, that something, most like a swift bird, had passed up the ditch, and one could not swear even to that. It seemed impossible that the flying something had been a four-legged animal.

It was the black rat. Nor did he go straight. He went, if I may so put it, every way at once, ending up with a merry-go-round dance with death--the harrier was pouncing savagely--round a tuft of grass, at such a speed that he looked exactly like the rim of a quickly spun bicycle-wheel--a halo, that is to say, and nothing else.

And then--he was crouched, panting, inside the hole, wondering whether his heart or his lungs would be the first to burst. And then? Oh, and then he--cleaned himself, naturally and of course. What else did you expect? He was the original black rat of Old England, and one of the cleanest animals on earth.

Mrs. Ratus, having vanished past finding while the hunt to the hole was on, presently scented her lord out, when the night had come and the harrier was gone, and together, starting like antelope at every hint of a sound, they traveled up the ditch, and up the bank of a stream that the ditch folded into.

Once an owl--the nomad, short-eared owl of the marshes--let forth a hoot that would have sent a nervous lady into "astericks," and sent *them* into nowhere, as if it had detonated a charge of that lively mystery called T.N.T. under their dainty feet. Once, just as they were lapping like dogs at the edge of the ice that was conspiring to span the brook, an otter shot up his head--jaws wide and dripping--almost under their long and pointed noses, and they, with one accord, and driven by their long tails acting as a spring, leapt simply into space. At any rate, they could not be followed by mortal eye, wherever they did leap to.

And once they met a wandering cat. And that cat seemed to go mad, for she shifted about the steep bank of that stream, and up, and about--here she swore because the spikes pricked her--and down a holly-bush, as if she had got a rocket tied to her tail. She had not, of course. She was hunting black rats. I suppose she saw them. If so, she was the only person who did, and I feel sure that, instant as she was, when she was up the bank or the holly-bush they were down it, and when she was down they were up. Finally, when her lost temper had completely run her out of breath, she slouched away, spitting like a worn sparking-plug, and very much disgusted. And--the black rats cleaned themselves!

That night was not a profitable one. The shell-fish of the estuary were gone, and there was little instead on the stream--only snow, and the snow fell quietly at intervals throughout the night, hiding everything. Rats, too, are creatures of warmth. They hate cold as much as the writer, and these two black rats became very miserable. They had no home, and did not know where to go; and, save for a few berries, they had nothing to eat. Mercifully, they had plenty to drink, and that is an item with rats, who die in a very few hours if they cannot get a drink.

A bitter, dull paling of the sky, which by courtesy we will call dawn, found them cleaning again, with their hand-like forepaws, exactly like cats, inside a water-vole burrow. The owner had been out, bark-chipping, all night--it was the only thing he could find to keep body and life from parting company--and was not over-pleased, on his return, to find that he had company at home. A short two-round contest ensued, during which the water-vole must have felt as if he had taken on a bit of black lightning. Then the water-vole went away, somewhat bewildered, to turn some smaller water-vole out of his winter bed; and the rats curled themselves up, heads between hindlegs, tail encircling all, with only their ever-ready, elfin ears poking out to give the alarm, and they slept. And, by the way, it was a saying in the wild that no one had ever seen them asleep, or knew if, or how, they did sleep.

Nothing came to disturb them during the day--which was a wonder, for all the wild was hungry and looking for food--and at the hour when

> Night, busy with her dawn, begins it with a star,

they came out, after a prolonged, starry-eyed stare, from their fastness, and continued their journey.

Things were serious now. They had not fed, and could find nothing to feed upon but two hawthorn-berries, dropped by the wasteful fieldfares; but they drank, *and* cleaned, and proceeded up-stream, with that caution one only learns in a world full of enemies and empty of friends.

Another six hours of this cold on an empty stomach would send them into that sleep--the dread, drugged slumber of the cold--from which there is no awakening in this world, and they seemed to know it. They were desperate, and their eyes burnt in their sharp heads like gimlet-holes of light. Desperate they were, as the poor, little, brilliantly resplendent, and tropic-looking kingfisher had, no doubt, been, whom they found, frozen into a dried, huddled heap, under the stream-bank, and so emaciated that, after they had picked his bones, they scarcely knew that they had touched him.

But anon the face of the snow changed--meaningly for them. Whereas before they had been alone, almost, in a frozen world, scarcely crossing a trail but the quadruple track of water-voles or the chain-pattern impression of a moorhen--nor had seen a living thing but the square-ended, squat, little, black form of a water-vole out upon an alder-branch, gnawing bark--they now began to be aware of gradually increasing company. Not that the company advertised itself, mark you. Being wild company, it would not; but they knew it was there.

The chain-trail of the moorhen reduplicated itself. It was joined by that of a water-rail--they saw his ruby eyes and rat-like form in passing. The fourfold track of a rabbit led the way ahead of them, as if pointing the path, to be joined by the broken footprints of another rabbit, and then by the track made by the longer leap of a hare, fourfold also. The delicate lined marks left by a wood-mouse now kept company with the others, and anon the little fairy imprints of two field-voles--short-tailed field-mice, if you prefer. They crossed the track of another rabbit going, at right-angles, down to the water to drink, and then the little, busy tattoo of bank-voles. Another hare's trail, and more rabbits' tracks, began to meander about, but all heading more or less one way--the way they were going. And then they stopped dead at the smudged groove and ancient and fish-like scent of an otter. Moreover, they had scarcely got over that than they came upon the dog-like tracks, and the smell, like nothing else, of Reynard, the fox; and, with nerves fairly tingling now, and eyes everywhere at once, they arrived at last--as the converging trails seemed to say they would--at the towering, smudged blur against the sky, which was the farm-buildings.

The black rat peered under the lower rung of a gate into a straw-yard, and heard the rustlings of little folk--field-vole, bank-vole, and wood-mouse--who had gone before him. There was no sign of the others; but that was not strange, for the hares and the rabbits had probably gone round to the kitchen-garden, for which they were making in their extremity of hunger; and the otter and the fox were, most likely, keeping each other off the fowlhouse.

Wherefore, plucking up courage, the black rat skipped into the yard, and made straight for the manger, where, in the inky blackness under the open-sided roofs, he could hear the long-drawn blowing and sigh of fat cattle lying down.

A pale moon came out behind him, and showed him tripping lightly over a bullock's broad back. Then he was up on the manger-edge, had paused to make sure, and was down in the manger, picking up crumbs and dust of linseed-cake and chaff. Three mice were doing the same thing, but fled at his approach; but he did not trouble about that, for the cattle had not left even him and his wife a full meal, having blown what was left of the chaff away, and licked up practically all of the cake-crumbs and dust. However, it was better than nothing.

The rat's natural curiosity was awakened, and his comparative warmth in this place, out of the razor-edged wind--oh, what a relief to be out of that infernal sawing blast!--made him explore. And he ran along the edge of the manger to a hole in the wall, which led--the peculiar and indescribable smell said so--through to the pig-sties. But here he stopped, and his wife behind him stopped. Some one was coming through from the opposite side--some one who smelt very much worse than any pig.

Next instant both black rats had gone off together like sparks--if ever sparks were black--and the brown rat, coming through the hole, wondered what on earth had happened. Then he sniffed at their trail, tried, but found it impossible, to follow, and passed on. He would have felt great pleasure in slaying them if he could, and they knew that.

The black rat now essayed to cross the yard to the stable. He could not very well stop there--up among the rafters, that is--all night, so he came down, and, with his wife following him, gingerly rustled out upon the partially snow-covered straw.

Then he got a shock that turned him into a winking series of black streaks.

Then he got another shock which turned him, literally, into--well, into black lightning. You never saw anything like it in all your life. You never would have believed that any living beast could have so frantically and so furiously got itself about from place to place so instantaneously. It was--dazzling. It made you blink. It was It in the agility line, and no mistake.

Firstly, the brown rat, having hidden up in some black corner, with brown-rat cunning, came hopping out instantly--nay, charging--on the black rat's trail. And there was murder in his wicked, little, glinting eyes at he came.

Secondly, a white eider-down quilt--at least, that was what it seemed like--descended lightly as--as an eider quilt, and as soundlessly, out of the blue-black sky, and covered the brown rat up. You could hear his horrid, muffled screaming of rage and fear under the quilt; you could see the quilt--but they saw that it looked pale brown on top--lifting about, and feeling for that murder-child of a rat underneath. Then the quilt got him--you could hear the unspeakably beastly death-squeal reverberate mufflingly--and then the quilt rose, still utterly without sound, and one saw it was a big barn-owl, with a rat--a brown rat--twitching in its white-mittened claws.

But do you think that made any difference? If so, you don't know the cruel devil of perseverance that is the brown rat.

As the black rat, at the end of his amazing lightning display, reached the barn, with his mate behind him, he leapt--he could not stop--clean over the back of one great twenty-inch, glitter-eyed brown ghoul, called by the death-scream of his colleague--other rats usually answer it--coming out of a hole. The black rats dashed into the hole like flickering streaks, but the brown rat had instantly spun upon himself, and was after them.

The barn was an unfortunate choice. It seemed full of brown rats, and four of them, in the darkness, instantly took up the pursuit of the now fairly hunted black couple. Nothing but their miraculous agility saved those two from being eaten alive, but they came out of the barn on to the spotless snow on the far side, with only a foot to spare between their long tails and the mangy, scarred head of the leading brown fiend behind them.

Straight across the open, like a drawn black bar, they shot, to a towering building of wood, and along this--here they lost six inches of the precious twelve by which they led--looking for a hole. They found it, whizzed in, five brown rats close behind them, nine brown rats hard behind the five.

They discovered themselves in a great room half-filled with sacks and the sweet smell of corn, and in and out among these sacks they led their hunters such a dizzy chase as no man ever witnessed, or could witness, for the matter of that, since human eyes could not follow it. But the end seemed positive, anyway. It was only a question of tiring the black sparks out, for the four brown rats in the place, engaged upon lowering the weight of the flour in the sacks--one of those rats a dreaded cannibal of twenty-one and a half inches--joined in the mobbing, and soon the black rats found themselves in such a position that there was no escape--no escape for any but a black rat. For them there was one way. And those two living electric sparks on four feet took it. They went up the wall!

I don't know, but I guess that, as the black rats' upper jaws were longer and sharper and more shark-like than the brown rats', and their tails very much longer, they got a spring off the tail--and legs, too--and had an agility in hanging on to knots and crannies above that possessed by the brown ghouls. Be that as it may, they did it, and got a respite under the floor of the room above, before their enemies, traveling more normally, and by holes, could swarm up after them.

Then the two, cornered at last, with one last desperate rush, shot up through a hole in the boards, out into the middle of the room on the first floor, and stopped dead.

Ah! they stopped. Good reason, too. Good reason had the five brown rats, excited with blood-lust, hard on their tails, to stop also.

They found themselves suddenly revealed in the middle of a big room, furnished mainly with a few sacks, and flooded with a dazzling, blinding glare of electric light, that seemed brighter than the very sun.

There they were, all seven, black and brown, struck rigid, plain and clear for any to see. And four men and two dogs stood there seeing them. They, those men and dogs, had just come quietly for their evening rat-hunt, turning on the light suddenly, for the place was a mill as well as a farm, making--from the mill-wheel--its own electricity.

There was a strained, aching pause for about as long as a man takes to gasp. Then the dogs sprang in, and one of the men jumped to the only hole in the room they had not previously stopped up.

But the black rats! The brown rats died, at intervals, fighting horribly, as cornered brown rats do. In five minutes they were, all five, dead--that is, all that had come into the room and been cut off. The black rats, however, in five minutes, were not dead. Nobody seems to have seen them, after the hunt had once begun, till the others were killed. Even then all four men aver that they could never rightly swear that they saw them. They saw lines, and streaks, and flashes, and whirls, and halos of black, which might have been rats--and the dogs said they were--but no one could swear to it. At times these giddy phenomena were among the rafters, at other times they were on the floor, and yet again they were going up or coming down the walls; but all the while both men and dogs seemed to be everlastingly too late, and hunting them where, half-a-second before, they had been. In fact, they perpetually had been, and were always where snapping jaws and beating sticks were not.

At the end of half-an-hour the men, mopping their foreheads, even in that cold, gasped, "Lor' love yer! Did yer ever see th' like?"

At the end of three-quarters of an hour the men flung themselves, gasping, on to the sacks of flour, and the dogs, panting, on to the floor--done. And the black rat and his mate, lively as ever, perkily watched them from the rafters.

Then the men and dogs went away, the light went out, and presently great sounds of war below suggested that the brown rats on the ground-floor were having the time of their lives. So were the two black rats, but a different sort of time. They were feasting upon meal and grain. And there, so far as I know, as they were like birds, flying among the rafters like black lightning if molested, they live to this day.

IX

LAWLESS LITTLE LOVE

She rolled over and regained her feet in a flash, to find herself facing a dark beast, with a huge, bushy, white tail, held up straight like a pleased cat's--but this was a sign of warning, not pleasure--that shone ghostily in the gloom of the mysterious, dread thorn-scrub. And the face of the beast was the face of a black and grinning devil, and its eyes shone red.

She stood there, shivering a little, with the tiny young thing crawling weakly away from almost under her feet, and the long, vivid, raw gash that the white-tailed beast, coming from nowhere special out of the night, had set upon her shoulder--a murderess caught in the act.

On three legs--her left hindleg had been bitten off by a trap set for a hyena--emaciated, with all her natural buoyant courage gone out of her, her wonderful agility gone too, she felt instantly in her heart that she could neither face this diabolical-faced foe, nor yet get away from it. This same crippled condition had spoilt her hunting forays, and, driven by hunger, had made her nose into other people's nurseries, and be caught just on the point of slaying somebody else's baby, when the owner had come home, like a streak out of the night.

But that was not the worst of it, for she was longer than the enemy, a bit, and might have put up a good fight--she had fought for her life, as a matter of course, ever since she left her mother's side--if the enemy had not brought with her an ally. It was not visible to the eye, that ally, but it was to the nose--a most distinct and appalling stink, and it could be felt, for it made her nostrils smart. Apparently, then, that white tail was intentional, was as a red flag, insolently displayed, warning all to beware of the stink. Well, there is more than one way of holding your own in the wild, and a most unholy smell is not such a bad way, either, when you come to think about it.

The owner of that nameless odor was a polecat--not our polecat; worse than that--and--well, you know the breed. Fear they know not; neither is pity with them a weakness, especially where the lives of their young are concerned. This one did not wait. She attacked quicker than you could cry "Knife," taking off with all four feet together, in a peculiar and patent way of her own.

The would-be murderess, who was long, and absurdly short in the leg, too, just like her opponent, only with a more graceful and not such a thick-set body, turned on herself in a snaky fashion, and her neck, that the fangs had aimed at, was not there when the polecat arrived, but her teeth were, and they closed on the polecat's cheek.

The latter gibbered horribly at the spark of pain, and set herself really down to fight.

The intending murderess said nothing at all, but, unbalanced with her game hindleg, having no force to push or spring with, and being very weak, she knew she was done for directly they closed to the clinch.

In a few seconds the polecat had her down, and only an awful, mad, desperate clashing of fang against fang kept the attacker off her throat. It could not last.

Then it was, at that moment, that a sharp little, gray little, dark-spotted, clean-cut, close-cropped, intelligent head, on a snaky, long neck, peeped out of the shadows, and peered about, as if to see what in whiskers all the pother was about. The head might have been there by chance, but it wasn't. Its owner had been running her trail for hours, and looking for it for days, and didn't mean to let her go, now that he had finally come up with her, polecats or no polecats, smells or not. But he was not a fool. He knew the game, the bitter, cruel game of death, as it is played throughout the wild. With man the inexorable law is, "Get on or get out." In the wild they phrase it another way: "Kill or be killed." Man puts it more politely, perhaps, but it's all the same old natural law, I guess.

The head and snaky neck developed a long, creamy, tawny-spotted body, and the body a long, banded, tapering tail--all set on legs so short, they scarcely kept the owner off the ground; and the name of that beast was genet. The same are a sort of distant relation of the cats, a fourth cousin once removed; but it is necessary to tell you, because you might think they were beautiful weasels, otherwise. *And she was a genet,* too--the murderess that might have been.

Then the new-comer moved. Then he began to move, and--here! It was just like the buzzing of a fly in a tumbler. Certainly you could say that he was still there, but you could not swear that you actually saw him.

The first that the polecat knew of him was that red-hot fork-like feeling that means fangs in the back of your neck. The polecat spun on herself, and bit, quick as an electric needle, at the spotted thing, that promptly ceased to be there, and, to use the professional term, she "made the stink" for all she was worth. She forgot all about the long female would-be slayer of her children, and the genet was mightily thankful to drag herself clear, but she would not have been she if she had failed to get her fangs home, as a parting shot, before she went.

Then, I fancy, she was ill; and, upon my soul, I don't wonder. It was enough to asphyxiate a whale-factory hand. But the male genet was not ill, or, if he was, he was moving from place to place too quickly to give the fact away; and by the time he shot up a tree, like a long, rippling, cream and tawny-dappled, banded line, he left that polecat considerably redder than when he found her, and weak, as if she had been bitten by leeches. The polecat had certainly saved her young, or thought she had, although I cannot swear that the female genet had really meant them harm; but she did not look as if she had saved much else. However, she held the field of battle, and the foe had fled, and that is supposed to be the sign of victory; but that had been done by her "gassing" methods, so to speak, not by fighting alone.

Rippling about among the branches, an incarnation of grace personified, and hunting for her by nose alone, for in the moonlight her exquisite creamy, dappled coat was invisible--a real piece of magic, this--the male genet quickly found her for whom he sought. She remained low, lying along a bough, line for line, shadow-patch for shadow-patch, flat as the very bark, and as undulating, until she felt sure that he would run over her; then she rose, spitting and snarling in his face, cat-like and vicious.

It was a poor kind of thanks for having saved her life, perhaps, but it was her way--*then*. And, anyway, who can blame her? She had never met any living creature that was not a foe or an armed "unbenevolent" neutral in all her life, and she did not know that any other category or creature existed, the recent fight notwithstanding.

But the male genet neither ran nor fought. He dodged her snap, by a tenth of an inch, almost without seeming to move, and there he stood looking at her meekly. She leapt to him, and he shot off, as she arrived upon, the place where he had been. Perhaps she knew that only a genet, or a mongoose, could do that trick in a manner at once so machine-like and precise; and after that she merely sat, bent in a curve, with her lips up. But her spring had given her away, and he saw that she was lame. Perhaps he saw, too, the gleam of hunger, the wild, cruel gleam that forgets all else, in her eyes; but who am I to say whether he understood it?

Be that as it may, the male vanished suddenly and without explanation, doubling on his trail and going out like a snuffed candle. He was in view, as a matter of fact, several times during the next few minutes, climbing quietly; but the dark blotches of the leaf-shadows magicked him into invisibility, and no one could tell where he was, till suddenly the silence was smitten by one piercing squawk somewhere among the greenery above. Then a crash, wild flutterings, a hectic commotion, and he and a terrified guinea-fowl came down together, more nearly falling than he liked. Indeed, he must have let it fall, or gone himself with it, as he slid past, grabbing for holds, if she had not dropped quickly to the next bough and taken a hold, too. Then, side by side, they hauled the warm, feathery, fluttering thing up, and he slew swiftly, in order to silence the noisy prey, who foolishly kicked up such a noise, as if maliciously; for he knew--and perhaps the gleany (guinea-fowl) did, too--how quickly a crowd may gather to interfere in an advertised "killing" in that wild.

The female genet, however, was past caring about risks. She had reached a stage of hunger when no risks can overshadow the risk of starvation, and she had the guinea-fowl by the throat, and was sucking its blood before the other had time to realize what she was at. Then, with fine discrimination, she ate the breast and thigh, and later might, or might not, have let him have a look in, if some blotched shape had not slid up, without sound, across the blue black night sky, and, halting in the tree, begun, apparently, to crack nuts very sharply and very quickly. Whereupon, without saying anything, the genets faded out.

It was nothing much, really--only the noise she makes when the giant eagle-owl is angry; but when you are a genet, with a body under two feet long, you may find it rather a bore, if nothing else, to remain cheek by jowl with an angry eagle-owl three feet or so across the wings, with the feline temper of an owl, and armed, owl-like, to the teeth, if I may so put it.

"An angry eagle-owl"—*Page 150*

[Illustration: "An angry eagle-owl"]

Now the question came as the two genets arrived at the ground--would she follow him, or would he have to follow her? He was determined, anyway, that nothing short of calamity should part them.

Yet I don't see, since he never uttered a sound, how she understood him to say that if she would follow him he would find her food, even though she was still hungry, for she had not yet got to trusting him much more than she trusted any other beast, and seemed to think that he was half as likely to eat her as to get her something to eat. Such a thing as another creature finding her food for only just friendliness--love was out of the question yet, or out of her question--was an idea her suspicious nature could not yet grasp. However, she followed.

Twining and twisting, turning and tripping, in and out among the bush and the tree-trunks, soundless, and quite invisible, except when they crossed a moonbeam--and then nearly so, because the moon has a trick of, as it were, dissolving the colors of even fairly conspicuous creatures--they crept on their low way. There was not a sound that they did not crouch for, often flat as a whip-lash--and that wild is full of sounds by night, too--not a puff of air that they did not throw up their sharp little muzzles to test, not a movement or the hint of a movement which their eyes did not fix with a suspicious stare.

They passed a hippopotamus feeding--a sheaf at a mouthful--upon long grass; they came upon three wild dogs eating an antelope and gibbering like gnomes; they beheld two striped zebras stampeding from a lion; they got into the middle of a herd of elephants--but what must those giants have seemed to them, almost at ground-level?--and did not know it, so silent can the mighty ones be, till they heard the unmistakable digestive rumblings; they happened on the tail of a leopard, observing a young waterbuck antelope, and retired therefrom without his suspecting them; they watched some bush-pigs rooting in a clearing, hoping they might turn up some insects worth eating; they heard a mother-lion grunting among some reeds, and were nearly run over by the stampede of zebras that followed; they chased a rat that ran into a hole in which was a snake, and it never came out again; they went up a tree after a weaver-bird's nest, but, from the way the bottle-shaped structure was hung, could not get at it; they investigated a hare's hole, and found a six-foot mamba snake, with four-minute death-fangs, in possession; they risked the thousand spikes of a thorn-bush to get at a red-necked pheasant roosting, only to find the branch he was on too slender to hold their weight; they were stalked by a wild cat, and hid in a hollow tree; and were pounced upon by a civet cat--who was their big cousin--and dodged him most wonderfully; and were chased by a jackal, whose nose they bit when it followed them into a hollow log. Finally, they came to the wall, and stopped.

Their noses told them it was not the wall of a native village, for no one, not even a man, could possibly make any mistake about *that*. Also, their noses may have told them other things. Anyway, the moon saw them, in the form of two gray lines, slide over the wall and drop silently into the shadow on the far side.

A wild cat was courting a domestic cat of the bungalow close by, at the corner of the compound, but, flat as strips of tawny-spotted cloth, they got past him all right.

A black-backed jackal was gnawing a bit of old hide at the angle of the wall, and they were forced to make a detour up to the veranda of the bungalow to avoid that sharp-eared, sharp-eyed one.

Here, on the veranda, they discovered a chair, and the male genet, standing on hindlegs to see what was in the chair, found himself looking straight into the electric-blue, purplish balls of light that betokened another cat, which had been asleep, but was now very wide awake.

He went round that chair in the form of a hazy, wavy, streak, as the cat shot out of it. The female genet faded from publicity behind a palm in a pot. But the genet's tail was so long that, with the cat and himself going round and round that chair like a living Catherine-wheel--both he and the cat spitting no end--the cat was touching his tail, while he was snapping at the cat's. Wherefore he moved across the veranda as an arrow flies, and round the corner, and as he turned the corner he--leapt.

It was a beautiful leap, and it cleared the danger that he seemed bound to run into, as it lifted in his path, by about an inch. As he sprang he heard the cat's claws scraping loudly, as she madly endeavored to stop--too late.

Then the head of the eight-foot python that had been creeping up round that corner in the process of stalking that cat whizzed by beneath him like a hurled poleax.

As he landed the genet heard the cat make one sound--only one--and it was indescribable, and he dropped off the veranda into the shadow of a bush, where the female genet presently joined him.

There was a small mongoose (my! what a lot of hunters do collect about the bungalows at night, to be sure!) under the bush, engaged in eating that precise reptilian form of poisoned death known as a night adder, which it had just killed. But the genets had other and private business, and they parted from the mongoose with no more than a snarl, the two genets to appear next--or, rather, to be no more than guessed at--crossing the last stretch of moonlight between them and the fowlhouse.

As they did so, a blurred, vast-winged, silent, dark shadow passed overhead, and a peculiarly piercing whistle stabbed dagger-like through the waiting, listening silence. Both genets jumped, as if the whistle had really been a dagger and had stabbed them, and vanished into hiding before the sound had ceased, almost. They knew that shadow--the owner of the whistle; they had met her earlier that night--the giant eagle-owl. But what the fangs and claws was she doing here? After rats, perhaps. They hoped so, and tried to think she was not after them.

The people who are condemned to live in those parts know that deaths, many and mysterious, go about there in the night, seeking victims, and that fowls must, in consequence, be well penned. Yet they die; and it has been said that where a snake can squeeze into a fowlhouse, there a genet can follow--perhaps dealing with the snake first, and the fowls afterwards. Certainly, there seems to be no longer, and narrower, and lower, and more sinuous little beast on this earth than the genet.

The male genet took the problem upon himself as his own special province to find entrance into places; and the female, her suspicions of him oozing away more and more every minute, "kept cave." And he found an entrance, that little, long, low beggar; he found an entrance, a hole up under the roof, that appeared small enough, in all conscience, to be overlooked by anybody.

The moon knows how they climbed to it--I don't. And as the male genet dropped down inside, the female took his place. But even as he landed he wished he had not. Fear was there before him.

In the smelly, stifling, heated pitch-darkness a fowl squawked with pain, and others burst into noise above his head.

Then he made a blunder. Surprised certainly, and angry perhaps, he growled.

Instantly the confusion ceased and hushed to silence; and instantly, too, round, large, amber-gold balls of light like lamps, to the number of two, were switched on--fixed upon him, staring, so that he "froze" in his tracks where he stood, and her crest stood up on the female genet, as it does on a cat, as she peered through the hole. They had disturbed something at its killing.

Very few graven images move less than those two pretty but small hunters did in the nest half-minute, while the fowls settled down again, and the genets tried--mainly with their noses--to find out what, in the wilderness or out of it, they had run up against this time.

At the end of that period there fell upon their stupefied ears the sound as if some one unseen were cracking nuts--nut after nut, very quickly--in the blackness, and both genets very nearly had a fit--a motionless one--on the spot.

Then they knew, most entirely did they know, and the knowledge gave them no end of a fright. It was the giant eagle-owl. She--it was a she--had beaten the robbers in hole-creeping, had outburgled the burglars, and outcrept the creepers, though goodness alone knows how.

The only difficulty was, who was going out first, and who alive, and who dead?

The male genet apparently knew about owls, and nothing of what he knew had shown that they were cowards. Nor was he a coward; but the wild hunters we not out to win the V.C., as a rule, I guess; and, if they were, he was not one of them. He was out to feed, not fight.

Possibly, while he was considering this, standing there with arched back--by reason of his long body and apologies for legs--in the darkness, the owl was considering the same thing. Anyway, both seemed to make up their minds in the same instant, and to act on it. Wherefore they arrived at the hole under the roof in the same instant, too; and you can take it from me that there are very few creatures indeed who can go into a hole, or come out of it, with such an amazing rush as the genet.

The result naturally was war, and red-hot at that.

Grappling, spitting, hissing, growling, snorting, coughing, the two fell in a heap to the ground--and an owl on the ground is one degree more of a spiked handful than an owl in the air--where they continued the discussion in a young whirlwind of their own, much to the perturbation of the roosting fowls, who woke up and added to the riot.

The female genet had gone out of the other end of the hole, like a cork out of a bottle, taking a scratch on the nose from the owl with her; but, finding nothing further happen, she now crept back and peered in. What she discovered did not give her any comfort, for, although upon her back, it looked as if that she-owl had been specially designed to fight that way. She had one fiend's claw gripped well home on the male genet's shoulder, and another doing its best to skin him alive; while her beak was hammering the gray top of his weasely-looking head. True, the male genet's fangs were buried up to the socket in the owl's throat, but that was no proof that he had found either her windpipe or the equally useful jugular vein, and, if he did not pretty quick, it looked as if it would never matter, so far as he was concerned.

I like to think of what that little, long, crippled female genet did then, in that well-like blackness and that smelly heat, with the chance of retreat open to her, and no one to say her nay. Without hesitation, she dropped to the ground beside the scuffle, and flung herself into it--into the winnowing, slapping radius of big pinions, that beat and beat and beat, smothering all with feathers and dust. One wing caught her squarely, and she fetched up against the wall, winded and dazed; but she was back again in a flash, dancing on her toes, and, suddenly flattening, shot in, level with the ground, like a snake.

She arrived. She felt feathers against her nose--she could not see. The wings pounded her flatter. She laid hold, biting in as deep and as far as she could get.

As a matter of fact, she had got the owl by the neck, but one would have thought she had turned on a young volcano by the confusion that followed. Both genets shut their precious eyes, and hung on, while that owl beat herself round and round in one last wild flurry, coughing horribly and humanly the while, and cracking nuts. Finally she collapsed as suddenly as a pricked bladder, and lay still--a great, mixed-up feathery heap, limp and pathetic, with her vast flung-out wings.

The two genets backed away, glad enough to be done with such a fiery, feathered fury. The male genet stumbled a little, and sat down. He was nearly as red as the sun on a stormy dawn, but all the blood was not his.

They did not seem to trouble further about the great foe lying beside them. Certainly she pervaded the air with a musty smell that was not attractive, or, at least, not attractive when fowls were by; and it was to the fowls they turned, the female first, the male later, after he had done some very necessary licking.

I fancy that, though dizzy, the male genet was rather proud of himself. He had brought his lady-love to such a feast as she may have dreamed of, and she had saved his life. That gave them a fellow-feeling that looked well for his prospects in love. But I do not think he had quite realized how hungry that beautiful velvet-skinned damsel of his choice was till that minute, and then he was given no time to think about it.

The dark over his head burst like a mine, and feathers and noise enveloped him whirling. That represented the female genet coming down, fixed to the throat of a hapless fowl. She sucked the blood, and flew at another. Ordinarily she would have removed that one and found it enough; but men who have been "broke," when they got suddenly rich, seem to go temporarily mad with the lust of spending, and so it was with her; only, her madness was the lust of killing.

She killed, leaping and wrenching at the poor, screaming birds' throats, blinded to the world with excitement, drunk with blood. That is an awful intoxication, and makes even men, let alone wild, carnivorous beasts, do unmentionable things. Also, the smell of blood was too much for the male genet, and he presently rushed, with flying tail, into the crimson orgy too.

They were some time at this craziness; and when they had finished, they and the fowls that were still alive could only lie and pant together among the contorted slain, the blood--you would never believe how a cockerel will bleed--and the carmine-tinted feathers. You might not believe me if I told you how many fowls they had killed, but it was a most disgraceful number, and quite inexcusable.

And then, even as they lay there, dead-beat, they started suddenly and together, for, almost like a blow, the fact dawned upon them that it was day. Night had stolen away, and dawn discovered them at the killing; and goodness alone knows how long they had been at it--ten minutes or hours. Anyway, here it was, and they leapt to their feet together.

As they hurried out they had to pass the place where the carcass of the owl *had been*! It was gone--mysteriously sauntered as a corpse into nowhere. Owls are uncanny creatures at any time, but moving about when dead is not usually a recognized habit of theirs. The genets sniffed anxiously, and ran the trail to the hole under the roof, since it happened to be on their way. Through the hole it went, and into the air--literally into the air. In other words, that owl had simply "bluffed" death when she realized that she was near death. The bluff had come off; and at a later, and what she judged a proper, time, she had just, and of course silently, flown off by the way she had come; and--as I live!--a fowl had gone with her.

One minute later an unsuspected martial hawk-eagle precipitated himself out of a big, hoary, old fig-tree, a hundred yards away from the fowlhouse, on to one of the genets' disappearing tails. This is the world's most general view of a genet, by the way--its disappearing tail; and it is given to very few to see the beautiful, dark-blotched, creamy, little, lithe, long beast that the ringed tail belongs to. Of course, the eagle was too late.

Two minutes later, a late leopard, returning to his lair after a blank night's hunt, saw the tail of the female genet, who was leading, disappear into a hollow tree. The male had not time to get in as the leopard sprang, so he shot up another tree close by, disturbing a mamba cobra, whose color was green, and whose bite was death, as it lay asleep among the twined vines. The legless terror fell to the ground and streaked for its hole, and the following leopard only just managed to bound out of its way as it did so.

Then, leaping light as thistle-down, coughing harshly, the leopard went up the tree after the male genet, and appeared to have cut him off from life and liberty for ever.

The genet climbed beautifully, and dodged round the tree-bole, and in and out among the trails and the leaf-bunches of matted creepers, with amazing speed; but the whole time the leopard's paw, all hooked claws bared, was whip-whip-whipping the air, only just behind that lovely, long, ringed tail of the genet, and more than once touched it.

Finally, hard driven, panting, at the end of his tether, it seemed, the genet was forced out upon a branch, farther and farther, slowly and more-slowly, the leopard creep, creep, creeping, almost flat, well spread and craftily, his paw, well out in front, hooking at the luckless little genet, till the twigs began to bend under the poor hunted creature, and all hope seemed gone from him, for the ground was sixteen feet below, and there was nothing between. And then--ah! but it was a fine effort!--just when it seemed that he could go no farther, and that the next terrible hooking round-arm stroke of the leopard must fish him into the annihilating scrunch of the terrible jaws, whose foul, hot breath already played upon him, the genet sprang.

It was a wonderful spring; the little beast had gathered every last ounce of his strength for it, and he literally seemed to sail out upon the air. Sixteen feet to the ground he bounded, and twenty-two feet out from the bole of the tree he landed, and--well, what d'you think of that?

Quick was the leopard--to our eyes he seemed to come down almost on the heels of the genet--but not quick enough, for he had first to gather himself on an uncertain, swaying footing. Wherefore, by the time he got to the ground, bounding like some great rubber ball, he had the pleasure of seeing the male genet's tail vanishing also into the small hole in the hollow tree.

And there he left them, because perforce he could do nothing else. And there, too, we leave them, curled up side by side in the darkness and safety, reconciled, and a happy couple at last.

X

THE KING'S SON

They found the king's son lying in a bed of reeds with his sister, the king's daughter, and although the prince and princess fought royally, as befitted their rank, they were smothered up roughly in sacks and carried speedily--the queen might return at any moment and want the captors--to the Governor of all the Provinces, and the Governor spake thus:

"Oho! A royal pair, eh? They shall be sent to the capital, but first we must put them in an inclosure while we knock up some kind of a cage."

And into "an inclosure" were they, therefore, cast, and it was small and bare, but for one box with dried grass in it; and the walls of the place were of corrugated iron nine feet high, so that escape looked impossible. Ransom was out of the question, and rescue a wild, but still faintly possible, dream--they could even then hear their father speaking in a mighty voice very far away, but their mother, they knew, would be following their trail in terrible silence.

Meantime they were king's children, and it behooved them to carry themselves as such in the presence of the enemy. Wherefore did they neither cry nor grieve (outwardly), nor sulk, nor cast themselves down or about with despair or rage. They just sat down side by side, and put their heads together, and stared with haughty insolence at the common crowd, "the lesser breeds without the law," who gathered to inspect them. It is not every day men get a chance to spit at and make mock of a king's son, whose father, as like as not, killed one's mother or little brother with no more thought than you or I would kill a rabbit, and the crowd made the most of that chance.

But luckily night, who was their godfather, came stalking swiftly westward, as he does in those wild parts, and flung his protecting cloak over them, and the crowd melted to its fleshpots, and the magic of the dark settled down over all.

One by one the little lights twinkled out in the huts and tents of their captors, and the deep bass drone of men's voices within mingled with the shrill cackle of women, and the high song of the mosquitoes without; and the smell of cooking and tobacco together came to them, so that they sniffed aloofly and stirred from their places.

A pariah dog, lean and yellow, came to eye them furtively through the chinks of the corrugated iron, and the horses snorted and stamped in their pickets, as the night breeze carried to them their scent.

Time passed, and the shrill voices of the women-folk ceased, the deep mutter of the men died gradually down, the lights faded, the scene was lit up only here and there by the sudden glow of a fire kicked into blaze by a sentry, but the song of the mosquitoes never ceased.

Then arose and uprose the strange, uncanny voices of the night, which, taken together, made up a background to the great silence which they seemed to accentuate. And the king's son bounded again. They were to him as a mighty call, those voices, from his own land--the land of the wilderness.

The rumbling thunder of his father's rage, breathing of death and destruction, had ceased now; but there were plenty more sounds, and the king's son, listening, knew them all. The distant "Qua-ha-ha!" of a troop of zebras going to drink; the peculiar snort of an impala antelope, scenting danger; the far-away drumming of hoofs of a startled herd of hartebeests; the bleat of an eland calf, pulled down by who knows what; the "Hoot-toot!" of a hippopotamus, going out to grass; the sudden shrill "Ya-ya-ya-ya!" of a black-backed jackal close at hand; the yarly, snarly whines of a hunting leopard; the snap of a crocodile's jaws, somewhere down in the nearby river; and, last, but by no means least in ghostliness, the awful rising "Who-oo!" followed by a sudden mad chorus of maniacal laughter, which told that somewhere a gathering of hyenas were--at their work!

The king's son was moving about the prison now, examining what he could see--especially of the walls--with his wonderful, proud eyes, and what he could not with whiskers and nose. He made no sound, of course--not so much as a whisper; and when his sister joined him, they were simply intangible, half-guessed shapes, drifting--there is no other word for it--through the gloom.

A man, even a colored one, might look long into that inclosure and, unless he caught the sudden smolder at the back of their eyes, never tell where they were. Indeed, the inclosure was in pitch-darkness itself, by reason of its high "tin" walls; and even when a weird yellow moon came and hung itself up to add to the general uncanniness of the scene, the prison of the king's son showed only like a well of ink.

Suddenly the silence and the voices of the crickets were broken into by the sound of scramblings by night. A nightjar fled from the tree overhead to the accompaniment of strange noises; and an unseen jackal, who had crept up to the very huts pessimistically, in search of anything awful, or offal, fled with a startled scurry. Apparently something with claws was trying to scrape away the corrugated iron.

Came then a scrawling scrape, and a thump. Then silence. But after a bit the noises began again--a fresh lot, and more violent. The pariah dog, who had come to investigate with his tail in the air, went away again, and quickly, with his tail between his legs; and in the same moment the king's son's head appeared over the top of the corrugated iron wall in silhouette against the staring, surprised moon.

Of course, and quite naturally, every sentry was asleep, or else even they could not have failed to realize that the sounds of desperate scratchings that followed were no ordinary phenomena, and might bear looking into.

Presently the king's son's body followed his head, and he sat for a moment, balancing clumsily on that narrow top, before vanishing suddenly, to the accompaniment of a heavy thump that was the last sound he made in the place.

Further and even more frantic scratchings followed, and anon the king's daughter, who certainly meant to die rather than be left alone in the hands of the foe, eclipsed the moon. A pause, and she, too, vanished downwards with a thump that was the last sound she made there in that place also.

A minute later, and she had joined her brother under the thorny guard of a mimosa.

For a moment or two the pair stood rigid as rock carvings, looking back, crouched a little, and deadly silent. Then the king's son turned and led the way to the river at a loping trot, and his sister followed in his tracks. They shook the dust--literally and daintily as a cat shakes dew from her feet--of the hated captors' fastness from their feet in little momentary halts as they went, and the place knew them no more.

But there is one point I should like to insert here. Go and try to climb over a corrugated iron wall nine feet high, and with nothing but the bare earth to take off from, and see how you succeed. Further, when doing it, remember that these royal children were so young as to be little more than babies. Then you may tell how they accomplished the feat. I do not know exactly to this day.

It is to be hoped that by this time everybody will be aware that the king's son and the king's daughter were lion cubs, the survivors, therefore the strongest and the fittest, of three lion cubs in a litter. It required, you will admit, some resource, courage, and intelligence to do what they had already done, considering their age; but the worst was to come. Having got out of the frying-pan, they must now face the fire, and be quick about it, too, if they didn't want to be traced and recaptured at dawn.

Arrived at the river's edge, they stopped and stared out across the dark swirls and unknown, cold depths. Lion cubs, as also the young of every other African animal, must assuredly be born with an instinctive and a very lively dread of all rivers and their occupants. Any horrible invention of death, they must have known, might be expected to lurk there, ready waiting for them in that underworld of dark waters; but if they felt fear, they never showed it, and the pride of their birth held true. Their hesitation was only momentary. The terror had to be faced, bravely or fearfully, as they would, but still faced, and bravely it was done.

Slowly and coolly the king's son waded out into the black, chill waters. He felt the current, which was strong, plucking like invisible great fingers at his legs; he felt the cold strike through tawny, spotted fur and skin to his belly, but he never looked back. His feet were whirled from under him; he trod upon nothing, cold, cold emptiness, and that was enough to terrify any grown beast, let alone a baby; but he struck out right manfully, and his fine eyes and face took on that regal expression of haughty determination that you see in the face only of King Leo himself and his mate, and in no other beast in the world. And the king's daughter unhesitatingly followed--a real princess, by gad, sirs!

Steadily the pair swam, heading instinctively, one presumes, up-stream, to counteract the drift of the ever-shouldering current. There was, perhaps, from two to five feet of water under their sturdy paws; but had it been twenty or thirty, I, personally, believe it would have made no difference. There were probably also other things under and around their sturdy paws--things very much worse and less innocent than any water--things they must have, dimly, at any rate, if not acutely, been aware of in an inbred sort of way; but they made no difference either.

"Make way for the king's son and the king's daughter. It is their will to cross the river. Hear, all of you loathly horrors, you lurking terrors--make way. Who dares check the will of the king's son?"

Once there came a mighty swirl on their right hand--or right paw, if you like--and the waters parted, with waves and the spouts of a geyser, to give up the monstrous nightmare head of a hippopotamus. Once something cold and leathery and ghastly touched the bottom of their padded feet; and once--but this was too awful for any expression by pen--something else, equally cold, but smooth, coiled, writhing, round the king's son's left hindleg, but providentially slid clear again, as he kicked like a budding International.

Then came the ordeal. It arrived in the shape of two knobs, that just were suddenly, and remained, motionless in the mocking moonlight, on the surface of the water. It might have been merely some projection from a half-sunken log. It might, but--well, there had grown in the air an unspeakable stench of musk, and that wasn't there before the knobs showed up.

Both lion cubs saw, both little royal ones smelt, and in some dim way, warned probably by a terrible knowledge handed down to them from their ancestors, both baby swimmers knew. Terror--real terror, of the white-livered, surrender-or-stampede-blindly-at-any-price kind--could never, it seems to me, come into those fine, regal eyes; but the nearest approach that was possible occurred in that instant, and they swam. Ah, how those infant lions swam! What had gone before was mere paddling; and whether or not they had ever swum before in their short lives--and I doubt it more than a little--there could be no question about them now; they swam like practiced hands, and almost as fast.

Followed a pause, terrifying enough in all conscience, and then, slowly, silently as a submarine's conning-tower goes under, so dived those knobs, and vanished almost, not quite, without a ripple.

The cool night-air showed the breath coming from the broad, brave, water-frilled cubs' heads in gasps. The silence gave away their frantic panting. You could literally see them straining every baby nerve and muscle, could note the jerks with which they fairly kicked themselves along. And the opposite bank, a black wall of bush and reeds, was very near now, yet far--oh, how far, to them!

Ssee-shhrr-r-rr-r-shrhh!

As a torpedo hurtles hissing along barely below the surface of the water, so hurtled the head--the head with its wicked eyes on knobs; the head with its vast, scaly, long snout, its raised nostrils at the tip, its shuddering array of jagged teeth, its awful, armed, diabolical aspect of conscious power--straight at the king's son. Without warning had it come, and with still less had it attacked.

Swim, oh, swim, little king's son, for your very life! But the king's son did not swim--at least, not in that sense. He turned. Yes, that is right--turned; and the monstrosity of the armed snout, that same being a crocodile, of course, was upon him even as he did so. There would have been no time to turn after--no life! Still, the king's son may not have known that. Maybe he turned, as a man attacked by a dog does, because he felt, in a cold, nervy sort of spasm all up his spine, the terrible defenselessness of his hind-limbs. And as he turned, he struck--bat-bat!--

struck with all his talons unsheathed; struck with every ounce and grain of power, and force of brain to back that power, in his system; struck as only a cornered cat can strike; struck like a--lion.

The result was astounding.

The crocodile had aimed, true to a hair--you bet, he being a croc.--to grab the king's son's hindlegs, and pull him under. He had not reckoned on the turn, and the turn did it. His snout struck hindlegs, which were not where they ought, by his calculations, to have been, but were four or five inches away to one side.

Quick as only a reptile can be, he canted, to remedy the error, but the impetus of his ten-foot bulk was still upon him; it carried him by. You cannot stop ten feet of bulk and five-feet-seven of girth of flesh and bone and muscle and armor-plates, going at Old Nick may know how many knots, in half-a-yard, you know; and it was the half-a-yard that did the trick.

The king's son was aware, as he half-rose and delivered that desperate blow, of a mighty bulk shooting by, of an overpowering, sickening stench of musk, and of eyes, through the foam and the water--two little, wicked, unspeakably cruel eyes on knobs.

His chance! And, quick as light, he took it. Ough!

The rest was chaos.

And that is about all, I think--unless you would like to know that their mother, the king's consort, who had been working grimly along on their trail since dusk, slid swiftly down the bank in that crisis, a fiery-eyed, long, gliding shape, and plunging into the watery inferno utterly recklessly, brought out, one by one, dripping, shivering, and by the scruff of the neck, first the king's son, then the king's daughter, and stopped not till she had placed them high up the bank, safe among the thorn-scrub, where they crouched together, side by side, listening to the cataclysmal threshings of the blind devil down in the black waters below there; and their father, the king, came up--pad-pad-pad-pad--behind them, to thunder out defiance at all the world above their sturdy, broad, intelligent heads, and purr his joy at their return. Moreover, he looked proud as he stood there in the moonlight, that royal beast; and I like to think it was not all looks either.

XI

THE HIGHWAYMAN OF THE MARSH

There was some sort of violent trouble going on down in the reeds beside the dike. The reed-buntings--some people might easily have mistaken them for sparrows, with their black heads and white mustaches--said so, swaying and balancing upon the bending reeds, and calling the makers of that trouble names in a harsh voice.

And all the rest of the reed-people were saying so, too. It was an amazing thing how full of wild-folk that apparently deserted reed-patch was. Each bit of the landscape, each typical portion, is a world of its own, with its special kind of population. This one produced unexpectedly a pair of sedge-warblers and a reed-warbler, atoms who gyrated and grated their annoyance; a willow-tit, who made needle-point rebukes; a water-rail, with a long beak and long legs, running away like a long-legged pullet; a moorhen very much concerned as to her nest; a big rat very much concerned as to the moorhen's nest, too, but in a different way; a grass snake, who glistened as if newly painted in the sun; and a spotted crake, who is even more of a running winged ventriloquial mystery than the corn-crake of our childhood's hay-times.

All of them were on thorns--though on reeds, really--and evidently highly rattled and in a state of nerves over the trouble in the reeds. And not much wonder either, for, judging by the sounds, murder was being done in there among the secret recesses of the swishing green stems--murder cruel and violent, in spite of the sunshine and the light of day.

And then, all of a sudden, in the midst of almost a gasp of silent horror, a moment of speechlessness on the part of the wild lookers-on, out came the trouble, rolling over and over and over upon the soft, short-cropped grass of the dike-bank, and--they all saw. Also, they all, except the little warblers, who were safe, more or less, and stayed to blaspheme the public nuisance, instantly and at the same moment remembered appointments elsewhere, and went to keep them with a haste that was noticeable and wonderful.

There before them was a hare. But a hare is a gentle and altogether negligible wild-person. This hare, however, was fighting, and fighting like several furies, and grunting, and making all sorts of unharelike motions and commotions against another beast; and that other beast was most emphatically not a hare.

It--or, rather, he--was big, as we count bigness among four-footed wild-folk in Britain to-day. Probably he could stretch the tape to twenty-three inches, of which about sixteen consisted of very long, low body, with sturdy, bear-like, dumpy legs, the rest being rather thick, furry tail; and, though nobody--without steel armor--might have cared to take on the job of weighing him alive, he would have turned the scale at about two pounds eight ounces, or perhaps a bit more. Not a big beast, you will say; but in the wild he ruled big, being snaky and of a fighting turn, so to say. He was, in fact, the very devil, and he looked it--hard as nickel-steel. A dull, tawny devil, with a peculiar purplish sheen in some lights--due to the longer hair--on his short, hard coat, turning black on his throat, legs, and tail--as if he had walked in black somewhere--and finished off with patches of creamy white on head and ears. There was an extraordinary air of hard, tough, cool, cruel, fighting-power and slow ferocity about the beast--a very natural born gladiator of the wild places.

Men called him polecat, apropos of nothing about him, apparently, for he had no connection with poles or cats; or foumart, apropos--and you wouldn't have needed to be told if you had got to leeward of him just then--of his *very* unroselike smell--that is, fou--foul, mart--marten, his nearest relation; or, again, fitchett, as our forefathers termed him.

But never mind. What's in a name, anyway? A big sloe-hare, with a leveret or two *not* for sale--and that doe's leverets must have been in the rushes somewhere--may, upon occasion, show unexpected fighting-powers. And this one did. The polecat was kicked in the stomach, and kicked and scratched in the ribs, and thumped on the nose, and kicked and scratched and thumped on the head, before he could get in the death-stroke, the terrible lightning-thrust at the brain's base, which, like the sword-stroke that ends the bull-fight, dropped the victim as if struck by electricity.

And then he had whirled, and darted headlong for the reeds. He galloped in an odd, jumpy, sidelong gallop, as if he were a sort of glorified wild dachshund.

It did not take him long to inspect the reed-patch, to search it from end to end with his nose. His mind was soon made up to the fact that the wretched leverets had vanished, and that no scenting of his keen nose could find them. They had gone, evidently, quitted, like the pair of obedient children that they were, while their mother was cleverly holding the foe, and making demonstrations in his front. And now the pair of them were probably far away, lost past all finding among the mazes of the fields. And there was nothing for him to do but go and dine upon the old hare, which he did, taking, according to his custom, little more than a bite and a sip before passing on.

Then he turned and meandered off on the war-path. And this was a serious business, and a busy one. It was downright hard labor, for he worked his ground properly and for all it was worth, having a lot to kill, and not much time to kill it in.

At times he sat bolt-upright, and stared knowingly around--because his short legs gave him such a limited view otherwise. At times he climbed a mole-heap. At other times he hunted head down--and again one noticed the hound-like manner--in every possible direction, questing, casting here, casting there, working back, throwing forward, describing circles, and poking into and out of every reed-patch, bramble-heap, furze-clump, or other bit of cover that that coverless land offered.

And then suddenly he stopped. And then suddenly he ran forward. And then suddenly, the scent carrying him right smack-bang out into the open, he dropped flat and began to crawl.

He crept and he crept and he crept across that absolutely bare, flat ground, with never a tuft of fur or a feather of a single live thing upon it to be seen, till one might have thought that he had gone mad, and was stalking an illusion--as many, not beasts, have done before him; only they were men, and blew their brains out--or went bankrupt instead--afterwards.

Finally he stopped. And this was the oddest thing of all, because, if any creature could show intense excitement without showing it--that is to say, without muscle, eyelid, hair, or limb moving--that polecat did then. And yet, stare as one would upon the absolutely bare grass, dotted only here and there with a stone, or a rat's skull, or a daisy--all looking alike in their whiteness from a distance--not a living, breathing thing was to be seen.

And yet there was a living, breathing thing there. Indeed, there were several. It was only the eye that was deceived, not the nose--at least, not the polecat's nose--by motionlessness.

Right in front of the polecat, within spring, still as the very ground, the huffish, whitish check of a peewit, lapwing, or green plover would have been mistaken for one of the many stones; the thin, curving top-knot for a stem of the thin, harsh grass: the low curve of the dark back, with its light reflections in green, for no more than the natural curve of the close-cropped turf. And she was on her nest, which was no nest, but a scrape, backing her natural assimilation with her surroundings to see her through.

And the polecat knew she was there, and he knew she was on her nest--she would not have been fool enough to keep there otherwise. His nose told him--not his eyes, I think, for that was nearly impossible. The thing was, he wanted her to move. That was what he was waiting for. No more than the twitch of an eyelid would do to show which end was head and which tail; for he could only smell her, and, in a manner, would have to pounce blind if she did not move.

She did not move, however, and in the end he had to pounce blind, anyhow, so suddenly, so unexpectedly, when it did come, shifting like a streak, that there seemed to have been no time to shout even, or gasp.

Yet that peewit found time in that fraction of a second to rise, open her wings, and get two feet into the air. And then the polecat took her, leaping with unexpected agility, and pulling her down out of limitless freedom and safety. There was just a rush, a snap, a wild-bird squawk, and down the pair went, to the accompaniment of furiously fluttering wings and in a cloud of feathers.

She had made a slight mistake in her calculations, that peewit, a matter of perhaps a quarter of a second, but enough. Nature has not got much room for those who commit slight mistakes in the wild, I guess; they mostly quit the stage before passing that habit on, or soon after.

For a moment there was a horrible, strenuous jumble of fur and feathers on the ground, and then the polecat's flat head rose up on his long neck out of the jumble, his eyes alight with a new look, and his lifted upper lip stained with a single little bright carmine spot. The peewit was dead.

He pivoted upon his shanks and examined the nest. It was empty.

He got to his feet with rapidity, and, in great excitement, dropped his head and began hunting. In a minute a mottled pebble seemed to get up under his nose and run. He snapped at it, and it fell upon the grass, stretching out slowly in death--a baby peewit.

He circled rapidly, stopped, swerved, and, at the canter, took up another scent. Suddenly, in a tussock of marram, his nose and he stopped dead. Nothing moved. Then he bit, and a second buff-and-black-mottled soft body stretched slowly out into the open as death took it--a second baby peewit. He circled again, fairly racing now, and so nearly fell over a third pebble come to life that it scuttled back between his legs; but he spun upon himself like a snake, and caught it ere it had gone a yard. He snapped again, held it, dropped it, and another downy, soft, warm chick thing straightened, horribly and pathetically, in the unpitying sun, and was still--a third baby peewit.

But there were no more. He hunted around and around for the next ten minutes, but never struck a trail. Evidently there had been only three. And all the time father-peewit, who had just come back from dinner, swooped, and stooped, and dived, and rocketed, and shot down and around, his wings humming through the still air above as he went clean mad, and seemed like to break his neck and the polecat's back in any and every one of his demented abysmal plunges, but somehow never did quite.

And all the time, also, the polecat, without seeming to take the slightest notice of him, was watching him out of the corner of his eye, waiting, hoping for a chance while he hunted.

But this was not intended as an exhibition of "frightfulness," though the beast had slain far more innocents than he could eat. It was part of his duty; and though men have accused his kind of being possessed of a joy of killing, the accusation is by no means proven. And, in any case, the accused might reply to civilization, "Same to you, sir, and many hundreds of times more so."

Anyway, he now picked up a young peewit and made for the nearest dike; then along this, and presently into the water and across to the other side, swimming strongly and well; then along a smaller dike, hugging the reeds as much as possible, and pursued by a running fire of abuse from the sedge and marsh and grasshopper warblers, from wagtails, meadow-pipits, reed-buntings, larks, and all the small-bird population of those parts, till he came to the sea-bank, called by the natives "sea-wall." This was a high, grass-bearded bank designed to constrain the waters of the estuary, and there, in a hole, curtained by a dandelion and guarded by the stiff spears of the coarse marram grass, he stuffed his victim.

The burrow was not empty when he came to it, for it already contained two moorhens' eggs; but there was still room for more, and one by one he fetched the remainder of his victims, mother and all, that way, and stuffed them into the burrow, with a plodding, steady, exact doggedness of purpose that was rather surprising in a mere wild beast who, if seen casually, would have appeared to the ordinary man to be merely aimlessly wandering about the landscape. And, mind you, this was not quite such a simple and "soft" job as it looked. Grit was needed to accomplish it, even.

There was, for instance, the sudden, far too suggestive, swirl in the water as he crossed the dike for the third time, loaded, that gave more than a hint of some unknown--and therefore the more sinister--haunter of those muddied depths of pollution, who took a more than passing interest in the smell of blood, and must, to judge by the swirl, have been too big to be safe. And that was probably a giant female eel, as dangerous a foe as any swimmer of his size--though he ate eels--might care to face. Then there was the marsh-harrier--and the same might have been a kind of owl if it wasn't a sort of hawk--who flapped up like some gigantic moth, and dogged his steps, only waiting--he felt sure of it--for the polecat to slip, or meet a foe, or have an accident, or something, before breaking its own avine neutrality. Then, too, there was the stoat, or, rather, not the stoat only, but the stoat and his wife, who would have murdered him if they had dared, and took to shadowing and watching him from cover in the most meaning sort of way. And, finally, there was the lean, nosing, sneaking dog, the egg-thief, who had no business there with his yolk-spattered, slobbering jaws, plundering the homes of the wild feathered ones--he who was only a tame slave, and a bad one at that. But the dog followed the polecat into a jungle-like reed fastness, and--almost never came out again! When he did, it was to the accompaniment of varied and assorted howls, and at about the biggest thing in the speed line he had ever evolved. He was no end glad to get out, and the distant haze swallowed him wonderfully quickly, still howling every yard of the way--for, mark you, that polecat's teeth, once felt, wore nothing to laugh at or forget.

These things he accomplished as the night was beginning to fall, and the solemn eye of the setting sun--such an eye of such a setting sun as the estuary alone knows; bloodshot, and in a sky asmoke as of cities burning--regarded him as he finished and stood back outside as one who considers. He was a grim figure of outlawry and rapine, alone there in that lonely place, amidst the gathering, dank gray of the marsh mists, the red rays touching his coat and turning it to deep purple, and his eyes to dull ruby flame; a beast, once seen, you would not forget, and could never mistake.

But his work was not yet done. He was hungry again, but for him there could be no more food yet, and he turned with the same immutable and dumbly dogged air that characterized so many of his actions, and made off down the "sea-bank." Once he hid--vanished utterly would better express it--to avoid the passage of an eel-spearer, an inhabitant of the estuary almost as amphibious and mysterious as himself. Once he very nearly caught a low-flying snipe as he leapt up at it while cutting low over the top of the "bank"; and once--here he sprang aside with a half-stifled snarl and every bristle erect--he was very nearly caught by a horrible steel-toothed trap, set there to entertain that same dog we have already met, by reason of the small matter of a late lamb or two that had suddenly developed bites, obviously not self-inflicted, in the night. Then he crossed the dike at the foot of the sea-wall, shook himself, sat down to scratch, and straightway hurled himself backwards and to one side, as something that resembled a javelin whizzed out of six straggling, upright, faded, tawny reeds at the water's edge, by which he had sat down. The javelin struck deep into the little circle of lightly-pressed-down grass where his haunches had rested, and he caught a glimpse, or only a half-glimpse, of weird onyx eyes, and heard strange and shuddery reptilian hissings. Eyes and noises might have belonged to a crocodile, or some huge lizard thing, or snapping turtle, but the javelin was clearly the property of no such horror, and was very obviously a beak--now, by the way, withdrawn.

Followed the harsh rattle and the swish of big feathers and vast wings--he felt the draught of them--the dim outline, as it were, of a ghost, of some great shape rising into the gloom, and as instantly vanishing over the sea-"wall," and he was alone, *and*--there were now three upright and faded reeds in the clump near which he had sat him down by the water's edge to scratch, *not* six. The Thing, the portent, the apparition, or whatever you like to call it, had been the *other* three; yet you could have sworn to the six reeds before it moved. And the worst of it was that he did not know from frogs and fresh water what the Thing was. He had never glimpsed such a sudden death before, and had no burning desire to do so again, for he was shrewd enough to know that, but for that fling-back of his, the javelin would have struck him, and struck him like a stuck pig, perhaps through the skull! Oh, polecat! The bird was a bittern, relation of the herons, only brown, and if not quite so long, made up for it in strength and fiery, highly developed courage.

Caution is not the polecat's trump card, as it is the cat's, but if ever he trod carefully, it was thence onwards, as he threaded the dike-cut and pool-dotted gloom. He came upon a lone bull bellowing, and gave him room. He came upon something unknown, but certainly not a lone bull, bellowing too; it was the bittern, and he gave that plenty of room. He came upon two moorhens, fighting as if to the death, but *he* was the death; and slew one of them from behind neatly, and had to go back with it, past both bellowings, to a second burrow in the sea-bank, where he put it; and later he came upon only seven great, mangy, old, stump-tailed, scarred, horrible ghouls of shore rats, all mobbing a wounded seagull--a herring-gull--with a broken wing.

The gull lived, but that was no fault of the polecat's, for she managed to run off into the surrounding darkness what time he was dealing warily but effectively with one of those yellow-toothed devils of murderous rats--whose bite is poison--in what dear, kind-hearted people might have said was a most praiseworthy rescue of the poor, dear, beautiful bird. (The poor, dear, beautiful bird, be it whispered, had herself swallowed a fat-cheeked and innocent-eyed baby rabbit whole that very day, before she was wounded; but never mind.)

The polecat, after one wary sniff, did not seem to think the rat worthy of a journey to the sea-bank and decent burial, and passed on, the richer for a drink of rat's blood, perhaps, but very hungry. He came upon a redshank's nest in a tuft of grass.

The redshank, who has much the cut of a snipe, plus red-orange legs, must have heard or seen him coming in the new, thin moonlight, and told all the marsh about it with a shrieking whistled, "Tyop! tyop!" But the nest contained four eggs, which the polecat took in lieu of anything bigger, carrying two--one journey for each--all the way to the sea-bank, to yet another hole he had previously scraped, or found, therein. One of the other two eggs he consumed himself, and was just making off with number four, when something came galloping over the marsh in the moonlight, splashing through the pools, and making, in that silence, no end of a row for a wild creature.

The polecat stood quite still, with his long back arched, his sturdy, short forepaws anchored tense, and his short, rounded ears alert, and watched it come, not because he wanted to, but because there did not happen to be any cover thereabouts, and to move might give him away.

When he saw that the beast was long and low, and short-legged and flat-beaded, his long outer fur began to bristle. Those outlines were the trade-marks of his own tribe--not his own species only--and were, he knew, more likely to mean tough trouble than anything else. Then he realized that the path of the new arrival would take it right towards him, and that was bad, because to move now and get out of the way was hopeless. Also, he could see the size of the beast now, and that was worse than bad--some ten inches to a foot worse.

The beast held a wild-duckling in its jaws, and the little body, with its stuck-out webbed feet, flapped and flopped dismally from side to side, as the animal cantered along with a somewhat shuffling, undulating gait. And then the polecat became transfixed. He had recognized the new-comer. He knew the breed, and would have given a lot not to have molested that redshank's abode and be found there.

The strange beast--palpably a large, sinuous, and wicked proposition--came right up to the polecat, standing there rigid, erect, motionless, and alone in the moonlight, with the fourth egg between his paws, and then stopped dead, almost touching him. Apparently, it saw him for the first time. Certainly it was not pleased; it said so under its breath, in a low growl.

The polecat said nothing, perhaps because he had nothing to say.

The beast was an otter, and an old one. Also, it appeared to be suffering from a "grouch."

The polecat felt uncomfortable. He was eyeing the other's throat, and marking just the place where he meant to take hold, if things came to the worst; but he knew all the time that the otter, although its eyes had never been removed for a fraction of a second off his face, was really watching the egg. The otter was a female; probably she had young to feed; the presence of the duckling darkly hinted at it. If so, so much the worse for the polecat.

Then the otter put down her duckling, and growled again; but the polecat might have been carved in unbarked oak for all the sign of life that he gave. Then--she sailed in.

It was really very neatly and prettily done, for, as an exponent of lithesome agility, the otter is--when the pine-marten is not by--certainly quite It. The polecat seemed to side-twist double, making some sort of lightning-play with his long neck and body as she came, and--he got his hold. Yes, he got his hold all right. The only thing was to stay there; for, as he was a polecat and a member of the great, the famous, weasel tribe, part of his fighting creed was to *stay there.*

When, however, hounds fail to puncture an otter's hide, any beast might be pardoned for losing its grip; but he did not. Between the tame hounds' fangs and his smaller wild ones was some difference--about the difference between our teeth and a savage's, multiplied once or twice; and the old she-otter, who had felt hounds' teeth in her life, realized the difference. Also, it hurt, and the polecat did not lose his hold.

Then, maddened, wild with rage, the rage of one who expects a walk-over and receives a bad jolt instead, that old she-otter really got to work. She recoiled like a coiled snake, and the polecat felt fire in one loin.

It looked like the contortions of one big, furry beast twisted with cramp, by the moonlight. You could not possibly separate the combatants, or tell that there were two. But the polecat only fought because he dared not expose his flank with the foe facing him. Now, however, as they both rolled he--

Hi! It was done in an instant. At a moment when the roll brought him on top, and when the otter was shifting her own hold for another, and more deadly, which might have "put him to sleep" forever, he miraculously twisted and writhed, eel-fashion, and with one mighty wrench--a good strip of his skin and fur had to go in that pull, but it couldn't be helped--he had broken the other's hold, leapt clear of the clinch, and was gone.

The otter was up before you could guess what had happened, and was drumming away on his heels; but she soon pulled up, realizing that a polecat may be slow in the books, but not so slow in real life, with her to assist speed. Anyway, she seemed slower; and, in any case, she could not hope to follow him in the intricacy of holes and cover he was sure to take to, like a fish to water. Moreover, she was spitting up blood, result of friend polecat's neat and natty strangle-hold on her throat, and felt more in need of the egg--which she had won, at any rate--than a wild-goose chase.

Like a thin, wavy line through the night, friend polecat betook himself to the sea-bank, to a hole in the sea-bank, to the very depths of that hole; and there, in the shape of two angrily smoldering, luminous orbs shining steadily through the pit-like dark, he stayed. Most of the time, I fancy, he used up in licking his wounds. They needed it, for, though clean, the punctures from the otter's canines had gone deep, and a red trail of drops marked the polecat's route to his lair--one of his lairs.

Not, be it noted, that he was entirely ignored. Blood-trails are always items of interest in the wild, especially in the dark hours while man sleeps. Thus there once came to the mouth of the hole scufflings, and the noise as of an eager, inquisitive crowd--rats, who hoped for a chance to get their own back on a detested foe. But one evil snarl from the wounded beast removed them, convinced that the time was not yet.

Once, also, something sniffed out of the stilly night, and that was a fox; but one snap from within, a perfectly abominable smell, and the narrowness of the accommodation proved too much for brer fox, and he, with an insolent cock of the brush, retired.

Then, too, there was a rabbit, not looking where he was going, who got half-way down the burrow before he realized the awful truth, and went out backwards, like a cat with a salmon-tin on its head.

But along towards dawn there came an altogether different sort of sound, somehow--a sort of a little chuckling sound; and the polecat, answering it, came out. He looked rather less awful now than when he had gone in. A form was standing outside--a dark, low, long form, like himself; and, like himself, you could easily become aware of it without seeing it, even with your handkerchief to your nose.

It was his wife, smaller, but no less dangerous, than he. She was carrying an old hen-redshank in her jaws, its long beak and one of its wings clearly silhouetted against the moon. And apparently she would be very pleased if her husband would come out of the hole and make room for her to stuff the redshank into it.

Then, together, they moved at their indescribable, undulating gait--they looked like a snake between them in the moonlight--along the sea-bank, till they came, with caution and many clever tricks of vanishing, in case anybody might be watching, to yet another burrow, screened completely and very neatly guarded by the splayed leaves of a bunch of frosted sea-holly.

Both beasts went into the burrow, at the end of which was a nest containing live things, which squirmed and made little, tiny, infant noises in the darkness. They were the polecats' children, four of them, all quite young, and all very hungry and very lively indeed; and they explained a good deal of the reason for the stores of food set by in other burrows in the "sea-wall."

But they did not explain quite all, for, unless Mrs. Polecat liked her dinner high--and there was nothing I could find in her methods to show that she did--or unless Mr. Polecat had got a craze for collecting specimens and eggs, or forgot where half of his trophies were hidden as a natural habit of absent-mindedness, one cannot quite see the reason for hiding so much so soon, before the young could feed upon the "specimens."

However, I suppose the two beasts knew their own business best. The old male polecat seemed to, anyway, for just as the first flicker of dawn was paling the eastern sky he went off down to the mist-hidden dike, and, in no more than ten short minutes, returned with an eel, protesting violently in that horrible way eels have, which he promptly proceeded to decapitate and eat.

The afternoon had still some little time to run, when the waving grass down the side of the sea-bank and the half of a glimpse of dull tawny gave away the male polecat leaving his "earth" for the war-path once more. Was he ever anything else than on the war-path if he moved abroad at all?

That, even from above, was, I swear, all the indication he gave of his exit. Now, although it is a rule in the wild that self-advertisement is most unhealthful, there may be times when a beast like the polecat may not advertise itself enough. And this was one of those times.

Far overhead, circling grandly on effortless, still, great pinions, swimming, one might say, in the dome of the sky, a big bird, known as a buzzard, was staring downwards with the flashing, sheathed glance of all birds of prey--and aviators--at the world below. She, too, had young, and simply had to find a meal. The hour was late, and her success *nil*. Perhaps that accounted for much. Perhaps, however, all she saw was that half-glimpse of dull, tawny fur, which accounted for still more; that is to say, she probably made a mistake.

Anyway, the polecat was suddenly aware of a sound like the swish of a lady's skirt in the air above him, and of a dimming of the light. He sprang forward first, and glanced up second--knowing the rules of the wild. But he was too late, for instantly the long, hooked talons of the bird came down through the grass, and gripped. It was an awful handshake, for the bird was a buzzard, we said, who is a sort of smaller and less kingly edition of the eagle, without the imperial power.

For a few seconds there followed an awful struggle--great wings beating mightily downwards, beak hammering, and fangs meeting the hammerings with audible clashings. It seemed that the bird could not quite lift the beast, and that the beast could not quite retain connection with solid earth.

And then the bird rose, slowly, strainingly, with her vast pinions winnowing the air with deep "how-hows!" Like mighty fans rose she, still gripping the struggling polecat hard by the back in a locked clutch of steel--up and up, and out over the estuary, growing slowly from a great bird to a medium-sized one, to a smaller, and a smaller, all the time fighting, it seemed, like a mad creature, to gain the upper air, to climb to the clouds, as a drowning man fights his way upwards in the water. And there was reason--the old polecat's jaws were fast shut in a vise-grip, as of a Yale lock, upon her throat.

Never a sound broke the silence that brooded forever--in spite of the wind--over the lake-like, flattened expanse of the estuary save the deep "how-how!" of the buzzard's superb pinions as she climbed slowly into the sublime vault of the heavens; never a sound from bird or from beast. The beast hung on, dumbly dogged, with fangs that met in the flesh beneath the stained feathers; and the blood of the bird mingled with the blood of the beast as it trickled slowly down over his mangled head, upon which one fearful claw of the buzzard was clutched in an awful grip.

The bird struggled dumbly also, upwards, ever upwards, gasping, with open beak and staring eyes, fighting vainly for the breath she could not draw, till at last the two were no more than a speck--one little, dark, indefinite speck, floating athwart the great, piled, fleecy mountains of the clouds.

And then, quite suddenly, so suddenly that it was almost like pricking a bladder, the end came. The magnificent, overshadowing pinions collapsed; the bird reeled, toppled for an instant in the void, and then slid back and down, faster and faster and faster, turning over and over, in one long, sickening dive back to earth.

"Turning over and over, in one long, sickening dive back to earth"—*Page 191*

[Illustration: "Turning over and over, in one long, sickening dive back to earth"]

A watcher, had there been one, might have seen, just as the last rays of the setting sun touched the steely reaches of the estuary, turning them to lakes of crimson, something, somebody--or bodies, truly, for they were locked together--suddenly appear, streaking down headlong from out the heavens. There followed a single terrific splash far out over the tide, an upheaval of waters, a succession of ripples hurrying outwards, ever outwards, to tell the tale, and then--nothing.

Next morning, as the sun rose, a party of mournfully shrieking black-backed, herring, common, and black-headed gulls were gathered around the soaked and bedraggled carcasses of a polecat and a buzzard, stranded by the falling tide upon a mud spur, and still locked savagely and implacably in death.

Half a mile away, in the darkness of her burrow, the she-polecat stirred uneasily in her sleep, and, waking for a moment, stared out at the still, silent, secret marshes, wondering, perhaps, why her mate had not returned.

And ten miles away, far up in their great nest among the boughs of a mighty Scotch fir, three downy, but already fierce-eyed, buzzard nestlings craned their necks upwards, calling hungrily, and wondering why their mother had not returned; while their father shot and swerved backwards and forwards over the tree-tops, mewing and calling, uneasily and lonelily, to the clouds for his wife, who had so mysteriously disappeared. And so--fate and the end.

This only remains to be said--the female polecat and the male buzzard did, in spite of Fate, manage to rear their young. And if the gamekeeper and the collector, the sportsman and the farmer, have not been too cruel, those young are alive-to-day.

XII

THE FURTIVE FEUD

There was a sun. You could not see it much because of burning, dancing haze, but you could not get anywhere without feeling it. Almost everything you touched--sand, rock, and such like--blistered you; and the vegetation, where it wasn't four-inch thorns and six-inch spikes and bloated cacti, was shriveled yellow-brown, like the color of a lion. Perhaps it was a lion, some of it. How could one tell?

Lizards, which were bad; and scorpions, which were worse; and snakes, which were worse than worse, lay about in the sun, as if they were pieces of leather drying. You could not see them--which was awkward, for some of them held a five-minute death up their sleeve--partly because they matched their surroundings, partly because they were still. They were colored burnt to hide in a burnt land.

Yet it was possible to be bright and gay and unobtrusive in this place, too--if you were cold-blooded enough not to boil dry and explode before getting a drink--for under some trees lay, in the old-gold, yellow, black-shade-streaked, tawny-red grass, a sleek and glistening, banded, blotched, and spotted, newly painted python. Yes, sirs, a python snake; and you couldn't see it in its new levee uniform--the old one lay not fifty yards away--any more than you could see the other, and plainly attired, bad dreams--so long as it did not move. Its length was not apparent, because it was coiled up; but it would have uncoiled out into something most alarming if stretched, I fancy.

The jackal made no sound as he came, tripping daintily, graceful and light as a rubber ball, into the scene, blissfully oblivious, apparently, of the fact that any other next step might awake a volcano under his feet.

He was a black-backed jackal; red-tawny sides, fading to nearly white under-parts; black back, grizzled with white hairs, neatly ruled off from the rest of him, like a big saddle; large, wide-awake ears; long, thin legs; bushy tail; very knowing eyes, and all complete--part wolf, part fox, and yet neither and something of both. No one living could, perhaps, have been agile enough to measure him, but he looked over two and a half feet from nose-tip to tail-root; and you can add, possibly, a third of that for the tail. But he was all there, whatever his length, every short hair of him, and none of the swarms of buzzing flies around seemed anxious to settle upon him.

He picked his way across to the shade of the trees, slouching quite casually, apparently; though how he avoided treading upon any of the sudden deaths variously thrown about seems a mystery. And just short of the shade of the trees he stopped. He had spotted, or scented--the latter is most likely, for the smell beat a chemical-works, a slaughter-house, and a whaleship rolled into one--the big snake.

The big snake remained motionless, and made no sign. Goodness knows whether it was asleep, if snakes ever do sleep. It certainly had its horrible eyes open, fixed in an evil stare at anything, or nothing, after the fashion of snakes, who are cursed in that they cannot shut their eyes to things. (Imagine the position of some people in this world if they were afflicted like the snakes!)

For about a minute that jackal stood like a carved beast in wood, with the original bark left on his back. Then he began to sink, slowly, gradually, till he lay as flat as a punctured bladder. And the picture of that little black-backed fellow--that *Canis mesomelas,* if you like official terms--all alone there, and surrounded by a dozen deaths at least, and all nasty, doing the stalking act upon that python was great. He stalked. My! how he stalked! And with reason, for he was taking on, perhaps, the biggest thing in the hunting line that he had ever tackled, and it was a million to one that, if he did not win, he died, and horribly, too; and he knew it. Ordinarily he would have been the python's prey.

"That little black-backed fellow doing the stalking act upon that python was great"—*Page 195*

[Illustration: "That little black-backed fellow doing the stalking act upon that python was great"]

There was a little snicker, as it were, in the air as his fangs closed, and the python, waking one-twentieth of a second too late, lifted its head. Then, short and crisp--snap!

Talk about tweaking a lightning-flash by its tail! It would have been a wake to what followed then.

The jackal knew what to expect--by instinct, I suppose. Anyway, he did not wait longer than it takes to scrunch as hard as possible with canine teeth as sharp as knives, and leap clear.

Ho did it, however, and stood well back, with his ears cocked and his head on one side. It was as if he were panting, "Now, let her rip"--and she did.

A hurricane in a cage, a volcano in an eligible house-lot, a geyser in a water-jug--what you will; but they were all tame alongside that python, after the little black-back had got his fangs home.

You know the size of pythons? 'Bout the biggest things in snakes there are going, bar two; and this one was not a baby. But nobody can properly measure their strength. This one unwrapped itself in one awful swiftness, and wrapped itself up again more awfully swiftly and in worse knots. Then things became hazy, and one could only tell by the dust, and the sand, and the grass, and the leaves, and the other things flying around that something was happening.

But the jackal did not seem to care. He only sat well back, with jaws open and very red tongue lolling, obviously doing a dog-laugh to himself. Perhaps it touched his sense of humor to think that so small a beast as he, with just one scientific bite, should create such a deal of disturbance. But the--er--aroma could not have amused even him, and he was, as you might say, salted to stenches; for, though he was on the up-wind side, even there it was enough to knock flat anything that the python's tail could not reach. It was a most stupendous stench--a sort of weapon of defense, or danger-signal, that these big snakes have.

Now, perhaps it was the reek that drew the purr. Purring is generally looked upon as a nice and comfy sort of a sound, but *this was not*.

The jackal just heard it intruding upon the confusion of the python's last contortions, as if suddenly, and it seemed to come from the ground, and the sky, and the surrounding scenery all at the same time. There was nothing nice and comfy about it at all. The jackal removed himself, at sound of it, about four yards in as many bounds, and every grizzled scrap of fur along his black back stood on end. If we had heard it, we should have reached for our rifle, and felt tingly all down our spine, for that was the sort of purr it was--a horrible, hungry, suggestive, cruel, and blood-curdling sound of ghoulish pleasure.

The jackal ceased to dog-laugh, and his tail was between his legs, for he knew that purr, and its name was death. Death angry is bad enough, but death pleased--

Louder and louder the purr became, till it seemed, as the python began to lash out the very last of its life, apparently, to fill the whole place. Finally, it became real, and--a shape walked slowly out of a thorn-bush.

It would be blatant exaggeration to call that shape a lion. It--he--had been one. He was now a walking hat-rack. Never have you seen such a lion. Never had the jackal seen such a lion, even; and he had done business--of the snatch-and-run order--with lions all his life.

However many years that lion had lived, to kill mercilessly, Heaven alone knows; but how on earth he had contrived to avoid for some time being equally mercilessly killed by hyenas, or wild hogs, Heaven and himself alone knew, too. He was a very, very old lion, a derelict of a lion, a shadow of a ghost of one, mangy, tottering, toothless--come down to eat snakes killed by others, even by jackals.

Then the jackal went away, dejected and disgusted. He was honestly proud of his slaying of that python.

It was the biggest screw-up of courage he had ever accomplished in his life, and to be done out of his rewarding big feast by that purring skeleton of a king of beasts! It was too much even for his pessimistic philosophy.

"Yaaa-ya-ya-ya-ya!" he howled, with his nose pointed to the brazen sun, and melted away among the accursed thorn-scrub with a look about him that said, as plainly as words, "And that's what comes of hunting in daylight."

The jackal, after a long skirmish and a drink, retired homewards towards sunset, when suddenly, from a tuft of grass ahead of him, a shadow shot and vanished. He picked up the trail at once, diagnosed it as that of a hare, and gave chase.

It was a fine chase, characterized by every aspect of first-class trailing, and carried along at such a speed that the quarry never got a chance to stop and get its second wind. Indeed, the quarry never had a chance to stop at all, until it was stopped, and the manner of that happening was strange.

Whether designedly driven, or whether by chance, one cannot tell, but the fact remains that the hare took a "line of country" which, if persisted in, would lead her close past the jackal's lair, or, rather, his wife's lair. This was important--for the jackal.

Once, indeed, our hunter all but overran a small--but quite big enough--boa-constrictor, which must have aimed to drop from a tree upon the hare passing below, and missed. It was in an even more evil temper, in consequence, than snakes usually are, and struck at the jackal with its head and shut mouth. The jackal quietly side-stepped, snapped, missed, and made off after his quarry, and about five hundred yards farther on he came up with "puss"--dead.

The jackal sat on his bushy tail, stuck out his fore-feet straight, and stopped as quickly as ever he could. Then he snarled, and full right had he to snarl.

The hare was lying on her back, weakly kicking out the last of her life with her hindlegs, and a stocky, short-nosed, evil, leering, side-striped jackal was standing over her. *He* had done the deed. And our black-back knew that side-stripe, had met him before. The two families lived only a few hundred yards apart, and it was Mrs. Side-stripe who was responsible for our friend's wife's crippled condition at that moment. This was a typical side-striper, one of the creeping, hunting-by-surprise-and-pounce sort, and it may be that he had never run down any prey worth speaking about in his life. In a way, he was the very opposite from our black-back, who was mostly legs, and a bit of a sportsman, and, I believe, really delighted in a good ringing hunt. Wherefore there was not much cause for surprise at the bitter blood-feud that had gradually grown up between them, till now things had come pretty well to a head.

The other beast folded back his lean upper-lip till his teeth glistened, and grinned at him--a menacing grin. I don't know if he guessed that it was, by all the laws of the chase, the black-back's hare, but he knew that he had pounced upon her as she passed--pounced like a cat, as was his way, what time he was profiting by his enemy's absence to keep that enemy's lame wife indoors, and from hunting even for insects or fruit, by prowling round her lair, and threatening her with growls. Perhaps he had designs upon her puppies. Perhaps his wife had. And perhaps Mrs. Mesomelas knew that. It is difficult to tell.

There was a sort of a blackish-tawny line drawn to the side-stripe--whose other and learned name was Adustus--and back. It scarcely seemed possible that the black-backed little chap had moved, but he had--leaped in and out again, chopping wickedly with a sword-like gleam of fangs as he did so. The other pivoted, quick as thought, and counter-slashed, and, before you could wink, Mesomelas was in and away, in and out, once, twice, and again. One bite sent a little flick of the other's brown fur a-flying; one missed, one got home, and the side-stripe's ugly snarling changed to a yap to say so.

Twice the two beasts whirled round and round, like roulette-balls, the black-back always on the outside, always doing the attacking, dancing as if on air, light as a gnat. Once he got right in, and the foe sprang at his throat. He was not there when the enemy's teeth closed, but his fangs were, and fang closed on fang, and the resulting tussle was not pretty to behold.

Mesomelas cleared himself from that scrunch with very red lips, but never stopped his whirling, light-cavalry form of attack. He was trying to tease the other into dashing after him, and giving up the advantage which his foe had in size and strength, but it was no good; and finally Adustus suddenly scurried into cover, redder than he had been, and our black-back, too, had to bolt for his hole, as an aardwolf, clumsy, hyena-like, and cowardly, but strong enough for them, scenting blood, came up to investigate.

Mercifully, the side-stripe seemed to attract the more attention, or shed the more blood, and while the aardwolf was sniffing at his hole--not intending to do anything if the jackal had a snap left in him, which he had, for the aardwolf possessed the heart of a sheep, really--the black-back managed to dash out and abscond to his hole with the hare. When the aardwolf came back, and sniffed out what he had done, he said things.

Our jackal's head appeared at his hole next dawn as a francolin began to call, and a gray lowrie--a mere shadow up among the branches--started to call out, "Go away! go away!" as if he were speaking to the retreating night. A gay, orange-colored bat came and hung up above the jackal's den--well out of reach, of course--and a ground-hornbill suddenly started his reverberating "Hoo-hoo-hoo-hoo!" and, behold--'twas dawn!

The jackal scuttled down to the river to have a drink, which he got rather riskily among the horns of drinking, congregated hartebeests, impala, and other antelope, and returned with the leg-bone of a bush-buck, which had been slain the night before by a leopard, and he went to ground very quickly, for the great spotted cat could be heard, grunting wrath, at his heels.

Then the day strode up, and the light, creeping in, showed our jackal, curled up and fast asleep, in his lair, as far away as he could possibly get in the space--two ant-bears', or aardvarks', holes run into one--from his also curled-up wife.

Later--for it was quite chilly--he came out to sleep in the sun, under a bush, till the sun, in turn, half-baked him, and he retired again to the den.

The days were, as a rule, for the jackal, a succession of sleeping blanks, but at the end of this day it was the fate of a small python--small for a python--to hunt a pangolin--who was as like a thin pineapple with a long tail, if you understand me, as it was like anything, or like a fir-cone many times enlarged, only it was an animal, and a weird one--into that den of thieves.

Mrs. Mesomelas, she appeared to shoot straight from dreamless slumber on to the pangolin's back in some wonderful way, and Mr. Mesomelas, he bounced from the arms of Morpheus into--the jaws of the snake? No, sirs; on to the nape of that snake's neck, if snakes may be said to have napes to their necks. But to get hold of the neck of a python is one thing, to keep there quite a different, and very risky, affair; and our jackal, who was no pup, knew that. If that legless creation of the devil could only have got his tail round something, our jackal might have been turned into food for his food, so to speak. Wherefore, possibly, he was frightened. It was like taking hold of a live wire by the loose end. Moreover, the space was confined, and there were the whelps and all, and I rather fancy black-back was more frightened to leave go and stay than he was to hold on and run.

Anyway, he held on and ran.

An old, fat zebra stallion, round-barreled and half-asleep, snorted suddenly, and stared with surprise at the sight of a black-backed jackal galloping as fast as circumstances would permit him, with the wide-mouthed head of a python in his jaws, and the remaining long, painted body trailing out behind. The snake was not going with any pleasure, and his wriggling tail was feeling for a hold every inch of the way, and if he could have got one--oh, jackal! But he could not, for the jackal kept on going, and the snake's after-length kept on trailing out straight, like a loose rope behind a boat, through the perishing glare and the heat-flurry that seemed to be making the whole world jump up and down, as it does when you look at it over the top of a locomotive-funnel.

Snakes take a long time to die, or to *seem* dead, even with a double set of glistening sharp teeth scrunching as hard as their owner knows how into their neck. At last, however, after a final series of efforts to get, and keep, in the shape of a letter S, the python's tail gradually ceased to feel for a hold, and the writhing strain in the jackal's jaws relaxed. Still, our Mesomelas was taking no chances, and he galloped home with his capture before he stopped, as proud and happy an old dog, rascally jackal as ever cracked a bone on a fine day.

He was a little puffed, and more than a little puffed up, and it may have been that he did not keep his eyes all round his head, as a jackal should always do. Anyway, there, in the gathering shadows of night, came a waiting, watching shadow, that was presently joined by another, and the two--their eyes glinted once in a nasty metallic fashion--stood head to head, watching him.

By the time Mrs. Mesomelas had hobbled out to view the "kill" for herself, and snarl her appreciation--truly, it was a strange way of showing it--with thin, wicked ears laid back, and more than wicked fangs bared, the waiting, watching shadows had crept forward a little, on their bellies, head up, and--Mrs. Mesomelas, with the quick suspicion of motherhood awake in her, saw them.

The snarl that she whipped out fetched the jackal round upon himself as if stung. Then he saw, and understood, and rage flamed into his intelligent, dog's eyes. It was the side-striped jackals, Mr. and Mrs., plotting to loot his "kill."

It was the black-back who attacked. Perhaps he knew that one secret of defense is swift and unexpected offense. Anyway, he attacked, sailing in with his dancy, chopping, in-and-out skirmishing methods; and Mrs. Mesomelas, on three legs and with the bill for the other to be settled, helped him.

It was very difficult, in the tropic dust, to follow what exactly happened next. For the next few minutes black-back was here, there, and everywhere, leaping and dodging in and out like a lambent flame. The human eye could scarcely follow him, but the human ear could hear plainly the nasty, dog-like snarling and the snap of teeth.

The side-stripe, as I have said, was the weightier beast, but the black-back never gave him the advantage, which he sought, of the close-fought fight.

More than once he was chased, but only to lead his foe into the open, where he could play his own game to his own liking; and at last, when the moon rose, and his mate had the female black-back driven back to her last ditch, so to speak, at the entrance to her lair, the side-stripped jackal, spouting blood at every joint, it seemed, collapsed suddenly, and apparently gave up the ghost.

Now, our black-backed jackal was not a young beast, and he was up to most wild-folks' games--which was as well. He approached the corpse with caution, and as he poised for the last spring the corpse was at his throat. Black-back, however, was not there, but his tail was, and the side-striped one got a mouthful of the bushy black tip of that. Whereupon

Mesomelas recoiled on himself, and for a moment a horrible "worry" followed, at the end of which the other dropped limply again, this time, apparently, really done for.

Very, very gingerly the black-back--himself a red and weird sight in the eye of the moon--approached, and seized and shook the foe, dropped him, and--again that foe was a leaping streak at his throat.

Mesomelas side-stepped, and neatly chopped--a terrible, wrenching bite--at his hindleg in passing. It fetched him over, and he lay still, the moon shining on his side, doubly and redly striped now.

This time it was Mesomelas who sprang at his throat--to be met by fangs. But in the quarter of an instant, changing his mind after he sprang, he shot clean up in the air, and came down to one side, and, rebounding like a ball, had the other by the neck.

For one instant he kept there, hung, wrenching ghastlily, then sprung clear, and, backing slowly, limping, growling horribly, flat-eared and beaten, the side-striped jackal began his slow, backward retreat into the heart of the nearly impenetrable thorns, where the winner was not such a fool as to follow him. And the black-backed jackal never saw him again. Living or dead, he faded out of our jackal's life forever.

And when he turned, his wife was standing at the entrance to the "earth" alone. The other, the female side-striped jackal's form, could be dimly seen dissolving into the night--on three legs.

"Yaaa-ya-ya-ya!" howled Mesomelas.

XIII

THE STORM PIRATE

The sea-birds were very happy along that terrible breaker-hewn coast. Puffin, guillemot, black guillemot, razorbill, cormorant, shag, fulmar petrel, storm petrel perhaps, kittiwake-gull, common gull, eider-duck, oyster-catcher, after their kind, had the great, cliff-piled, inlet-studded, rock-dotted stretch of coast practically to themselves--to themselves in their thousands. Their only shadow was the herring-gulls, and the herring-gulls, being amateur, not professional, pirates, were too clumsy to worry too much.

Then came the rain-shower. Not that there was anything in that. Rain-showers came to that land as easily as blushes *used* to do to maidens' cheeks--rain-showers, and sudden squalls, and all manner of swift storm phenomena. But behind the rain-shower, or in it, maybe--it blotted out cliff and inlet and sandbar and heather-covered hills, and, with the wind, whipped the sea into spume like an egg-whisk--came he, the storm pirate.

A guillemot--you know the guillemot, the fish-hunter, who flies under the waters more easily than she flies the air above the waters--had risen, and was making inshore with a full catch, when the squall caught her without warning. For a little she faced it, her wings whirring madly, her body suspended in mid-air, but she not making headway one inch against the sudden fury of a forty-mile-an-hour wind. Then, since she could no longer see the shore, which was blotted out with hissing rain, she turned and ran downwind, like a drawn streak, to the lee of a big stack of rock.

The next that was seen of her, she was heading out to sea at top speed, in wake of the rain-shower and the squall, which had passed as suddenly as it had come; and behind her, pursuing her with a relentless fury that made one gasp, shot another and a strange bird-shape. Its lines were the lines of the true pirate; its wings long and sharp-cut; its beak wickedly hooked at the tip; its claws curved, for no gentle purpose, at the end of its webbed feet; its eye fierce and haughty; its uniform the color of the very stormcloud that had just passed--dun and smoked cream below, and sooty above. True, he was not big, being only twenty-one inches--two inches less than the herring-gull. But what is size, anyway? It was the fire that counted, the ferocity, the "devil," the armament, and the appalling speed. Just as a professional boxer of any size can lay out any mere hulking hooligan, so this bird carried about him the stamp of the professional fighter that could lay out anything there in that scene that he chose--almost.

The guillemot flew as never in all her life had she flown before, and every known artifice of dodging she had heard of she tried, and--it all failed. The terrible new bird gained all the time steadily, following her as if towed by an invisible string, till at last he was above her, his wonderful wild scream was ringing in her ears, his cruel eyes glaring into hers, his beak snapping in her very face, his claws a-clutch.

No, thank you. In sheer terror she opened her beak and dropped her fish. It fell like a column of silver, and in a flash her pursuer was gone--nay, was not gone; had turned, rather, into a second column, a sooty one, falling like a thunderbolt, till he overtook even the falling fish, and wonderfully snatched it up in his hooked bill ere ever it could touch the waves, without a word or explanation of any kind whatever.

That, apparently, was his manner of getting his living; a strange manner, a peculiar way--the way of the pirate on the ocean.

Despoiled, but safe, the guillemot rattled away "for another cast"; but the foe settled, riding lightly on the lift and fall of the bottle-green waves.

Here he was no longer a wonderful phantom spirit of the storm, but just a bird that might have been passed over at first glance as simply a seagull. But not at a second glance.

Men called this strange bird Richardson's skua, or Arctic skua, or lesser skua, or, officially, *Stercorarius crepidatus,* or, most unofficially, in the vernacular, "boatswain," or "man-o'-war," or "gull-tormentor." Apparently you could take your choice what you called him. But he did not belong to Mr. Richardson really. He belonged to nobody, only to himself, to the wind and the rain, that seemed to have begot him, and to the grim north, from which he took his other name. He might have claimed the gulls as his near relations--they loathed him enough.

For a long time he sat on the lifting, breathing swell, floating idly. There was nothing else on the face of the lonely waters except himself and a flock, or fleet, I should say, of razorbills and guillemots, very far away, who alternately showed all white breasts, and vanished--as they dived and rose all together--like white-faced, disappearing targets, and one gull, who wheeled and wheeled in the middle distance, with one eye on the divers and one on the skua, as if, gull-like, waiting on a chance from either.

Then at last the skua rose again, and swept hurriedly out to sea to meet a small black-and-white speck that was coming in. It was a little, rotund, parrot-beaked puffin, loaded with fish--sprats--four of them set crossways in his wonderful bill. He seemed to know nothing about the skua till that worthy was upon him, and then, as he fled, after a furious chase of about three minutes, he suddenly surrendered by letting fall all his spoil.

The skua caught up one sprat before it hit the surface, but, being too late to overtake the rest, seemed to take no further notice of them, but swept on, to settle upon the water a mile away and preen himself. And this was where the waiting, watching gull came in--the herring-gull. He sprang to strenuous life, and, arriving swiftly at full speed over the spot, snatched up off the surface, and by clumsily attempting to plunge, two more of the sprats, before the skua could intervene.

Then it was that a terrible and a totally unexpected thing happened, and yet, if one comes to think about it and study the matter more, the most natural in the world; probably, also, on those wild seas, even common-place. Only, you see, there was no interval at all between the skua sitting placidly on the lap of the waves, eyeing the gull vengefully, and that same skua shooting straight upwards, all doubled up, on the top of what appeared to have been a submarine mine in a mild form in active demonstration.

"Shooting straight upwards on the top of what appeared to have been a submarine mine in a mild form"—*Page 209*

[Illustration: "Shooting straight upwards on the top of what appeared to have been a submarine mine in a mild form"]

This submarine mine, however, in addition to the burst and heave of torn and upflung falling waters and foam, had a visible heart, a great, shining, wet, torpedo-shaped body, which rose on end beneath the stricken bird, and fell again with a splintering crash that shot up the heads of the diving birds half a mile away. It might have been a thresher-shark, or some other northern shark, or it might have been a dolphin, which is bad, or a killer whale, which is a good deal worse, if it had not been a great gray seal seeking dinner; and its effect on the luckless skua was the effect of a battering ram, and the skua that fell back again with the fall of snarling water was to all intents and purposes a corpse.

But it was a good thing that he was so. Had it been otherwise, had he tried to get away or fluttered, there would have been no more of him. That is to say, the head of the seal came up--or its wet and suggestive big nose did--and poked about, trying to find the bird. It had evidently meant to grab him, to engulf him utterly and forever in the first rush; but something--some unlooked-for lift of a wave or turn of the bird--had made the shot miss, or nearly miss, so that the bird had been hit by the bloated six-footer's nose, instead of being crushed in its teeth--its terrible long and glistening array of murderous teeth.

All the same, the nose blow was bad enough. It was like being hit by the beak of a torpedo at full speed, fit almost to bash a boat in.

The seal was quite evidently looking for the bird, and, equally quite evidently, seemed bound to find him. To know why it did not at once see him is to know that the seal's view, from below the surface, of the world above is about a twelve-foot circle of white-gold light, that is all; and the skua, floating limp and floppy, had been, by chance, till then always carried hither and yon by the waves just outside that circle. But that chance could not last.

Then came the other seal. Came she easily and gracefully, as a seal should in her element, effortlessly gliding along, her head from time to time up like a dog's--some gentle dog's, say a mild-eyed spaniel's--looking about. She was just a female seal. She knew nothing of the bird or her companion, who were at sea-level, and more often than not hidden in the trough, till she came sliding down the slope of a round-barreled swell, practically on top of them. Then it was too late to avoid mutual recognition.

Quick as sound she had seen, had realized, had spun on her apology for a tail, and had gone, leaving a little trail of foam behind her to prove her speed and her coyness. But, quick as light, the magnificent male seal had sunk from sight, leaving a little chain of bursting bubbles behind to mark his speed. And the last that was seen of that lady seal was a speck far on the horizon, going like a masterless torpedo, alternately leaping forward through the air and shooting along on, or just under, the surface--switchbacking, they call it; and that, I dare to fancy, if it proves anything, proves that the coyness was only make-believe, and that she had allowed the daring admirer to catch her up and force her to act as if she were already vanquished and using the last arts of swift swimming she knew.

It left the skua, however; left him still floating, floating, floating up one long breaker's side, and sliding down its other side to its fellow behind, towards the shore--always towards the shore. It is true that the tide was falling offshore, but that made no difference to the currents of those parts, which were independent currents and of a great force. They were shouldering the skua steadily to land, and if you had dropped a line overboard there, with an ordinary lead on, you would have felt them pulling at it, and taking the lead along like a live thing. And the currents were Fate, so far as that bird was concerned.

There was a little inlet, and a little bay in the inlet no larger than a good-sized dining-table, and seaweed, green and red, upon the rock-bowlders that encircled it, and old-gold patches of sand between the rock-bowlders, and green grass behind the rock-bowlders, and brown-plush furze behind the green grass, and a patch of blue sky over all. And in the middle of the little bay in the inlet, bob-bobbing on the lap-lapping of the littlest waves, that--sifted out by then, as it were--had found their way so far, floated the skua, the Richardson's or Arctic skua, dead, to all appearances, as the proverbial door-nail. But that was not the rub. The rub was in the--ah!

"He-oh!" pealed down the clear, ringing bugle-cry from above, and a shadow floated upon the reflections a-dance on the surface in the little bay in the little inlet--floated and hung, so that it exactly covered the skua like a funeral pall; floated, and hung, and came down. As its claws scraped a bowlder, and it furled its long, narrow vans, it was revealed as the big herring-gull--him we left out upon the face of the waters, watching and waiting on chance.

His spotless expanse of head and neck alone marked him, gave him away, a speck you could see for a mile. His size--just on two feet--proved what his snowy hood proclaimed, in case there were any doubts. A smaller gull, an uncommon common gull--of eighteen inches--came and looked, to make quite sure--and went away again. The herring-gull, in spite of his silly name, has a reputation, and a "plug ugly" one.

And the herring-gull, he--did nothing. That is the strength of the herring-gull--doing nothing. He can do it for an hour, half a morning, or most of a day. His battle-cry might well have been, "Wait and see," but he must be one of the few living, breathing things on this earth who have made the game pay, and--lived. He might have been a lump of chalk, or a marble carving, or a stuffed specimen, or asleep, or dead, for all the signs of living that he gave. One began to wonder if he ever would move again. He had been a bird, but was now the life-size model of one cut in alabaster, with clear pebbles for eyes--they were quite as hard and cold as that, his eyes.

And all the time the body of the skua floated, and danced, and drifted, and lifted in, making an inch on one wavelet, to lose three-quarters of it on the next, but still, unnoticeably perhaps, but undoubtedly, gradually, surely, for all that, drifting in.

Somebody has written somewhere that gulls never touch a carcass on land. Sometimes they touch things on land which were not carcasses before they touched them. This gull, however, did not wait for any landing. Perhaps he knew that, once stranded, the gray crows might come to assist him in their own peculiar way, or a raven, who would not assist him at all, except into the next world, if he did not relinquish all claim to the feast. Anyway, whatever we poor mortals may kid ourselves into thinking he did or did not know, or what we may think he ought to have known, he began operations as soon as the skua came alongside, so to speak--that is, drifted against the particular bowlder upon which the sphinx-like herring-gull happened by chance--always by chance, of course--to be standing.

Now, there is no particular joy in having your eyes hammered at by a blunted sharp instrument, like a herring-gull's beak, for instance, even if those eyes happen to be shut, as I think the skua's were, and the instrument wielded with the extreme clumsiness of the half-trained, as I know the herring-gull's beak was. But, all the same, it was the kindest thing that could have happened, for, had it not been for that, the skua was like to have drifted in that fashion from that little inlet out upon another sea; not the one connected with the inlet, but one where you can drift forever, and whose name is Death. The physical pain, however, brought him round. He was only stunned, and the agony of the eyes, or eye, rather, was acute.

He opened the other eye--a wonderful, piercing, fierce orb. He contracted his feathers. The world grew from a mist in that eye to a little bay in a little inlet, with the seaweed-covered bowlder-rocks, the old-gold sand, the green grass, the brown-plush furze, and the patch of wonderful blue sky over-top. Then it took in the spotless, gaunt form of the herring-gull, and--he remembered that he was a skua, only some twenty-one inches long, 'tis true, but still a skua, to be treated and respected as such.

Wherefore, who so surprised as that big father of herring-gulls when the bedraggled, smoky-hued thing under his bill, which he may, or may not, have taken for a corpse, woke up, returned to life suddenly, and erupted into his very face, with the yells of a fiend, the weapons of a fury, and the rage of several devils? He yelled, too, that herring-gull, not entirely with rage, and did his best to get under way as quickly as might be, but became, before he knew where he was, altogether too busy even for that.

Not being in the habit of performing optical operations upon Arctic skuas as a rule, he had nothing in his memory to warn him of what followed, nothing to put him up to the absolutely diabolical fury of the onslaught he had to meet in the next few seconds. He certainly did his level best with such weapons as Nature had given him, but his blunt, hooked beak and the claws he had not got seemed suddenly meager against the hammering, tearing, stabbing, rending dagger weapons of his--"meal."

The skua saw red, and the herring-gull saw mainly red skua, as he was hurled back and down under the first rush, and instantly, without a second to recover, was hurled, equally helplessly, the other way, shrieking for his very life, and decorating the air and the old-gold sand with a pretty little cloud of his spotless feathers.

He fought with the desperation of almost all the wild-people, when there is no help for it; hammering, too, but wildly and clumsily. The skua fought with the cunning and precision of the professional, plus such a rage as can only be described as berserk.

It was not a long fight. Perhaps the skua felt that, after his collision at sea, his bolt would be soon shot, and he had better do what he was going to do as quickly and thoroughly as possible. Certainly he did appear to do so; and when at length he drew back, rocking and gasping like a drunken creature, the famed purity of that herring-gull's uniform was a thing of beauty no longer. That part of him, indeed, which was not red was mud, or sand, or green slime, and in his eyes was the most worried and tired look you ever saw.

He rocked, too, in his gait, as he ran and blundered, and he gasped with his beak open. When he rose, which he did without any sort of procrastination, he rocked worse than ever, and twice nearly fell, and once hit the water, before finally slowly dragging himself away upward, flapping low and heavily across the little waves.

With the one available eye--the eye left him in working condition by the herring-gull's clumsy efforts--flaming like a live-coal, that implacable skua watched him go. He may or may not have known it, of course, but I feel pretty certain that he would a few thousand times rather have been standing there upon the old-gold sand, with only one eye doing duty and an unspeakable agony in the other eye, than be that herring-gull in the condition he was then, going back to the bosom of his tribe. It is not a thing to dwell upon in polite society, but I tell you that the gull-folk do not always treat their wounded well, and there would be no chance, no earthly chance at all, of his finding a place in all that vast horizon of sea and sky and island where they, the ceaseless, never-resting "White Patrol," would not eventually find him.

Then the skua lay down, and thereafter surrendered himself to that utter reaction which birds, who live more intensely in action than almost any other creatures, have brought to an apparently exaggerated pitch. He did not sleep, but he did not move, and every muscle in him, every fiber, every nerve, faculty, organ, was surrendered utterly to rest.

Night came fluttering her sable wings across the scene, breathing and sighing audibly in the first silence that wild landscape of storms and squalls had known throughout the day, and the skua moved. His neck went up straight, and his head turned, looking sharply this way and that, fierce apprehension written upon him.

There was nothing one could see to give cause for this. A flock of curlew were passing, wailing one to the other, across the sunset; a string of late gulls trailed athwart the sky; and a wedge of those beautiful little wild-duck known as wigeon was letting itself down to the shores of the inlet. Far out to sea a black line, which might have been a sea-serpent if it hadn't been scoter ducks, trailed undulating over the waves, and a single great white gannet plunged from aloft into the deep at intervals with a report like a sunset gun. But they were all innocent, except in the opinion of the fish and shell-fish, and no manner of folk to trouble the pirate skua. Set a thief to catch a thief, however. And, besides, there was blood on the bird and around him, or the taint of it, and blood is the devil and all in the wild. There was nothing to be *seen*. No. That was the worst part of it. It was what was unseen that the skua was thinking about.

Wherefore, then, our friend of the pirate rig rose and walked stiffly to the summit of one of the bowlder-rocks right at the water's edge. He was by no means recovered yet, or in any condition for a fight in that desolate scene, and had to select the most strategic position he could crawl to. He did, and awaited Fate's reply.

The day died, and the moon came out to wink and dodge and play a foolish game of hide-and-seek in and out among the clouds. She showed the skua, a black knob atop of the black blob of his bowlder, apparently fast asleep, invisible if we did not know he was there. She showed black dots bobbing upon silver lanes, which were sea-duck of various kinds--scaup, long tail, scoter, and the rest. She showed a line of old, rotten posts, broken off short by the waves, along a sand-ridge, which were wild-geese; and she showed three big, white swans--wild-swans, wilder even than the geese--floating like ghosts in the enchanted light.

But she also showed other things, indistinctly, 'tis true; but enough--quite enough. She revealed for an instant, as she shone on the spot on the sand where the skua had sat, the fact that the sand seemed to be alive, horribly alive, as if the pebbles had taken legs and ran about. It was a sudden, ghastly flashlight, hidden as soon as seen, and it gave one the shudders. Those pebbles were crabs mad with hunger, as crabs always seem to he.

They had arrived there as if by magic--been creeping in ever since dusk, probably (one of the things that were unseen); but whether blood, or feathers, or taint of blood, or what horrible, ghoulish system of espionage drew them, is not for me to say. They were there, anyway, and--and--well, and then they were not there. The next flashlight of the moon showed that some others had taken their place. This was ghastly, for the others were bigger than any shore crabs, and they hopped, and they sat up hunched, like hobgoblins, and--they scratched! This last identified them, for the soulless, shelled crab-people are not given to scratch much--at least, not in *that* way. They were rats--shore rats. The last designation is necessary, for there are rats *and* rats, all bad, but the shore rat is the worst. How many sleeping birds, wounded, tired, or unalert, die at his hands, or, rather, his teeth, in the course of a year would amaze anybody if known, and the shell-fish he relieves of life are legion.

The hard, horny carapace of a retreating crab scraped, in the dead silence, against the rock-bowlder on which the skua sat. He made no move at the sound, the suggestive sound; but his feathers were shut down quite tight, and he looked far smaller than usual. When birds shut down their feathers in that fashion they put on an armor coat, as it were, through which very little can pierce. It showed that he was ready.

And you think that the mere shore crabs could be nothing to him. But a few hundred ravening shore crabs, with their lives for sale--all digging pieces out of you in the dark--are not so easy a proposition to dispose of as people may think. Try it.

One of the rats turned suddenly and faced towards him. The skua could see its little, cruel eyes gleaming like gimlet-holes in the wall of a lighted room. Then another, and another, and another did the same.

The skua was scarcely bleeding at all now, but he had left enough of a trail for *them*--they who make a specialty of the job. And they followed it. Hopping grotesquely across the mottled, hurrying patches of moonlight they came, one behind the other, and without noise.

The skua remained as still as the bowlder he sat upon. In that position, even peering closely, you would never have seen him, unless, like ourselves, you knew he was there. But he was drawn together, drawn in all his muscles like a tense spring, and--though this his persecutors could not know--he was recovering from his hurts rapidly, with the wonderful power of recuperation of all the wild-folk, who pay their price for it in clean, hard living.

Then suddenly there was a scuffle below him in the dark. One of the rats squeaked a little, acknowledging receipt of a crab's pincers closed upon him, or her. Followed the sounds of some scuttering, confusion, and the horrible slide and scrape of horny shells upon stone. Then silence, and the skua knew that, in that wonderful way they have, the crabs, at any rate, were gone--for the moment.

Remained, however, the rats, and one peered up over the bowlder the next instant, its eyes glinting in a momentary splash of moonlight fiendishly. Also, his quick ears could hear the soft creepings of the others on every side of the bowlder, back and front. They had surrounded him, and, like wolves, would now rush, and then--and then---- They had gone.

Yes, there could be no shadow of doubt about it. There had come an instant's furtive, hurried movement, a glimpse--no, half a glimpse--of hunched forms hopping through the dark, and they were no more.

The skua stared, and as he stared a great terror seized him. What more deadly form of death than themselves had they suddenly become aware of, to cause them to invite themselves into nowhere in that magic fashion?

In the dead silence that fell, he could hear nothing, see nothing. Yet he felt--indeed he knew--something seemed to tell him, that a deadly foe was at hand.

Hours passed. They were minutes really, but they seemed hours to him. Nothing happened; nothing showed. But the rats did not come back. Therefore, whatever incarnation of death it was that removed them must be there still. He knew that. That lonely, wounded bird knew that. And he was right.

Behind him, practically invisible, flat to the ground, a long, low, narrow, dark shape was lying crouched, creeping, creeping, creeping towards his tail. Slowly, almost painfully slowly, it drew upon him gradually, so gradually that the distance between them could scarce be seen to lessen. And soundlessly, so soundlessly that even his quick ears, trained far beyond the quickest human aural perception, could not hear it.

Then, so quickly that the eye could not follow it, the crouching form made its rush.

The skua was sitting motionless, with his head looking straight in front of him. The dark form came from behind, and there would have been no time for the skua to move before the thing, whatever it was, had him by the back of the neck, and dead, save for one little tiny fact. As it propelled itself forward, in the first bound, the claws of the beast's hindpaw's scraped upon a stone. It was only a little sound, and it gave the skua barely a fraction of a second's warning; but, he being a wild thing, it was enough.

Quick as light the bird had half turned upon one side, and flung up one claw and wing to cover his neck, whilst his head jerked round hindpart before in the same atom of time.

Thus it happened that the beast, unable to stop, found himself with his head and eyes being dug at by a hooked beak, and his jaws closed upon a skinny leg instead of upon the skua's spinal column, as he had intended, which would have put the skua out of life like turning out a gas-jet.

And it was then, in that instant, that the moon chose to dodge from behind a cloud and reveal the beast as a big, long, lean, and hungry dog-stoat. Probably he had thought that the skua was a gull, and a wounded one. There is a difference, however, between the skuas and the gulls, though they bear a family likeness. He discovered the difference now, and for the next few minutes was not overjoyed at the knowledge.

One cannot do much blood-sucking to weaken one's prey out of a scrawny leg that resembles a twig wrapped round with leather. And the stoat found this out, too, and he would have shifted his hold to the bird's body like a flash, if he had been given a chance, but he never was.

Before he knew what was happening, he was blinded by the beating of vast wings, his claws began to slip and slide, and--oh, horror!--still slipping and sliding, he found the bowlder going from him. It went from him, receding downwards with terrifying rapidity, and the dancing, silvery, sparkling water was sliding below, too.

Being a stoat, he hung on with V.C. doggedness and courage; but it was the worst thing he could have done. Moreover, as it was, he forced the bird to attempt reprisals in mid-air--a terrible proceeding.

Now, this was difficult, might almost seem impossible; but the skua is one of the most wonderful flyers that haunt the seas even--and most of the best are there--and what he could not execute in the air was scarcely worth mentioning. It included in this case a perfectly diabolical scraping of the foe's head with his available claw, and after that, since the dogged stoat did not leave go, and the pain was excruciating, a wonderful bend forward, and, at a pronounced and dangerous angle, a fiendish stabbing at the stoat's head with his murderous beak. This last involved a drop of nearly a hundred feet, but it did the trick.

Blinded, dazed, shaken, and maddened by the agony of blows upon his sensitive nose, the stoat opened his jaws to grip higher up the leg; and in an instant he was gone, turning over and over, down, down, down to the hungry waves below.

Ten minutes later the skua was calmly and safely asleep upon the top of a frowning black stack of rock, untroubled, I think, even by dreams of the terrible things he had gone through.

Next morning, an apparition of wonder and fierce beauty, the skua, quite recovered except that he had a lameness in one leg and a weakness in one wonderful eye that would last him a lifetime, came racing downshore out of a stormcloud into the full gold of the sun at some seventy miles an hour. He was in pursuit of a common gull who, with more luck than judgment, had caught a fish.

The gull held on for a few minutes, on and in and around the horizon, going like the wind up and down and around, as for his life, with friend skua ever close to his tail, before a wild yell, which he could not mistake, sounded in his ear, and he dropped the prize. The skua executed his wonderful dive, and caught the gleaming silver thing before it reached the waves, and shooting up again, was just about to continue his course, when a constant and peculiar flickering above the beach caught his telescopic eye.

He checked, flung up, came round beautifully effortless, and headed towards the sight. Probably he knew what it was, had fathomed it even from that distance. It was a gang of gulls flying round and mobbing a hapless wounded gull on the beach.

It was a foul thing to do, a horrible, blundering, clumsy murder, done slowly; but even so, it was all over before, with a scream that rang like the battle-cry of a Highland chief, and set the murderous heads up in wild alarm, the skua came shooting, twisting, turning, diving, and darting right into the heart of the crowd.

And they went circling, and wheeling, and hurling downwind like sheets of paper, those murderous sea-birds, dispersed and scattered over the face of the waters, and were gone almost without a word.

Then the skua came lightly down, rocking on the wind, and settled beside the poor, draggled, white body, no longer white, upon the shingle, which had been so foully done to death by gulls of various clans. He may, or may not, have known it, but I can tell *you* that the gull was the self-same herring-gull who had tried to kill him the day before. Now he--but we will draw a curtain here.

Next day the skua went away, and the fishing wild-folk breathed a sigh of relief as they watched him go, and for three days peace brooded over the winged fishers of those parts, so that birds could feed upon what they caught, nor be in fear of getting hunted for it. But upon the fourth day the skua came back. And he was not alone. A dusky, nearly brown--for they vary much in color--female skua came with him. And in due course they built them a home on the ground among the heather, and they guarded their treasured eggs and reared with amazing fierce devotion their beloved young.

Before his advent that strip of wild sea-coast had been, mercifully, without its skuas. Our bold buccaneer, however, having won his footing, took care to see that, so far as one bird could accomplish the great task, it never should be again.

XIV

WHEN NIGHTS WERE COLD

And the Northern Lights come down
To dance on the houseless snow;
And God, who clears the grounding berg,
And steers the grinding flow,
He hears the cry of the little kit-fox
And the lemming on the snow.--RUDYARD KIPLING.

A snipe rose suddenly, and began to call out; a capercailzie lofted all at once, with a great rush of winged bulk, above the snow-bound forest; and a white hare slid, like a wraith of the winter, down a silent forest aisle.

Then came the White Wolf of the Frozen Wastes, the terror of the blizzard, ghost-like, enormous, and swift. In dead, grim silence came he, loping his loose, tireless wolf's lope, and stopped at a windfall, where two forest giants, their decaying strength discovered by the extra weight of snow, lay prone, one across the other.

For a moment he paused, nose up, testing the still, cold air; then he leapt upon the upper fallen tree. He had, seen up there and clearly, an enormously thick and woolly coat, that magnified his already record size. You see him trotting along the tree-trunk. Then he stopped and stared down into the dark hollow under the upper tree. Then he sniffed--audibly. Then he licked his nose--and very red was his tongue. Then--but this he couldn't help, I verily believe, as he balanced there with his pricked ears and bright eyes--he whined.

And instantly his little, impatient, dog-like whine was answered by a deep, deep growl, that seemed to come out of the bowels of the earth.

He was just in time, as he leapt lightly off the windfall, to avoid the rush of a vast brown bulk, reeking of carrion, furry, terrible, with live-coals for eyes, and threshing the air with claws Heaven knows how long, which hurled itself like an avalanche out of the hollow at him. And that thing was a bear.

Now, bears do not, as a rule, without extraordinary reason, in that land, rouse themselves out of their winter sleep for the mere whine of a wolf. They are impregnable where they are, and know it. The extraordinary reason, however, was present. The white wolf was sniffing at it now--the bear's blood-trail to the windfall. Bruin had been roused once before that day--by beaters. He had then been driven forward, shot at by hunters, wounded, escaped, and returned to his den. But--but, I give you my word, if those beaters, those peasants, had known the White Wolf of the Frozen Waste was out, nothing in the wide world would have induced them to beat for bears or anything else in that vicinity.

The white wolf stretched himself to a canter, and slid away through the forest, dropping the trees past him like telegraph-poles past a railway-carriage window. He looked like the very spirit of winter, the demon of the snows, and stood for that in the ignorant minds of the sparsely scattered people--perhaps because at a short distance he was nearly invisible. His white coat, which was simply a conspicuous curse to him when there was no snow--which was one reason, maybe, why he retired from the limelight to some lonely fastness during summer--was an incalculable asset to him in winter, and he knew it.

He ran, with his smooth, loose, effortless lope, perhaps a quarter of a mile, then stopped, and putting up his head, howled a howl so full of hopeless, cruel yearning, so vibrant with desolation, that it sounded like the cry of a soul doomed forever to seek something it could never find. It was a lugubrious yowl there, in that setting, and it made one's scalp creep all over one's cranium.

And instantly almost, even as the last, long, horrible echoes died, sobbing adown the blue-haze perspective of the forest glades, the answer came--a far-away, fluttering, wandering howl, like the moan of the wind in its sleep.

The white wolf waited a moment, then howled again, and the ghastly sound came back to him, louder and nearer this time.

A third time he howled, and the forest cringed under the reply.

Then at last the shadows between the ranked tree-trunks took unto themselves life, and eyes, eyes in pairs, horribly hungry, cruel port-holes of brains, with a glary, stary, green light behind them, suddenly appeared everywhere, like swiftly-turned-on electric lamps. There was a whispering rush, as if giants were swiftly dealing cards in the silence, and--the White Wolf of the Frozen Waste was away, racing like a cloud-shadow, rapid and impetuous as a greyhound, at the head of a pack of one hundred and twenty-nine empty-stomached wolves.

They made no sound as they tore, compact as a Zulu impi, over the spotless white, because they had no trail to follow, only this huge devil of a leader; and they had their work cut out to follow him, for he was the longest-legged male wolf any of them had ever set eyes upon.

Straight as a twelve-inch shell the white wolf headed back to the fallen trees and the bear's den. When he reached them, he stopped so abruptly that the wolves behind him almost sat on their haunches in the effort to pull up. Those that failed fell sideways under a rain of wicked snaps from him, that followed one another quick as the stutter of a machine-gun.

The pack did not stop--at least, not the flanks of it. They swept on without a pause, out and round, like flood-water round a knoll, joined at the far side, and--were still. As a maneuver, a military maneuver, swift, unexpected, faultless, and silent, it was perfect.

For as long as a man takes to light a pipe there was dead silence, broken only by the quick motor-like panting of the pack. And one hundred and twenty-nine pairs of eyes regarded the fallen trees.

Then the white wolf, all alone, with hackles up, stepped forward and leaped upon the trunk of the tree that was poised upon its fellow. He ran lightly up it till he was exactly above the hollow formed by the junction of the two trees, then stopped, looking down.

Half-a-dozen of the older and more cunning wolves followed him; the pack surged forward until both trees became lined with a row of wolves, without breaking the circle of the main pack outside, and then stopped. All this in silence.

Then--but you could almost hear the trees breathe while he did it--the white wolf yawned very deliberately, and whined, insolently and very audibly.

The answer was instant.

Something rumbled within the den, deep down, like a geyser.

The white wolf whined again, and sprang aside just as the bear, maddened with the pain of a .450-caliber rifle bullet in his stomach, and seeking a sacrifice, hurled out of the dark and up over the tree-trunk, striking, with appalling nail-strokes, right and left; and the quickness of those strokes was only a less astonishment than the agility of the wolves getting out of the way of them. But--but he had come out to abolish one wolf, that bear; not one hundred and twenty-nine.

The white wolf dropped without a sound upon the bear's great, broad back. The half-dozen old wolves followed him like figures moved by a single lever. The pack sucked in with the rush of a waterspout. The bear vanished under a wave of fangs and tails, as a sinking boat vanishes beneath the billows. And the rest was the most diabolical devils' riot that ears ever heard.

The bear unwounded, even if he had been induced to come out at all, might have fought his way home again; but the bear wounded and cut off was a different matter. He battled as only a cornered bear can battle;

but the exertion of it gave the .450 bullet its chance, and he died--horribly--as they die who are pulled down by the starving wolf-pack, and that is not printable at all.

He took three wolves--smashed-in-heads and chests--with him to the other world, that bear, and left three others well on the road there. *All six* followed him by the path he had gone when the pack had done with him; but the losses might not improve the temper of the pack, though they partially stayed the hunger of a few. And the white wolf seemed to know that. Full devilish indeed was the cunning of that brute.

Scarcely was the last bone cracked, scarcely the last wisp of skin snapped up, than the white wolf, wet, and red and wringing over the head, was away again, at full speed--and his full speed was a thing to gasp over--with a wild and rousing howl that gave the pack no time to ponder on its casualties.

This time also there was no trail, so the pack had full leisure to concentrate all its energies upon the job in hand, or paw, I should say--namely, galloping. No, racing would be the only word; for the white wolf, knowing his kind, perhaps, gave the pack no leisure to grow dangerous over its losses or its hunger. Only idleness gives time for questions to be asked about leadership, and he kept them busy; and if they wanted to keep up with him at all, they must needs extend themselves to the full.

Soon he led them to a clearing running, straight as a railway cutting, through the forest. Out in the clearing, he dropped his head, howled, flung half round, and began to follow tracks; but the scent was enough for him without the tracks. They were the footprints, the sleigh-trail, and the hoofprints of the beaters, the hunters, and the pack-horses, loaded with game from the hunt of the day that had just gone.

For a moment the pack, even *that* pack, his pack, the pack of the White Wolf of the Frozen Wastes, checked a little, shied, and were dumb. They were used to his leading them upon some hectic murder-raids, but never one quite so blatant as this.

Quickly, however, the real pain in their empty stomachs got the better of them, and they swept round and began to follow--half-a-dozen here and there--with whimpers. And then, the excitement spreading, they all rushed in, and breaking out, with a blood-curdling rush, into the full-throated chorus, "Yi-yi-i-ki-yi!" of the wolf-pack in full cry--an M.F.H. who had never heard wolves might have mistaken it for the music of a pack of hounds if he had listened to it from a distance--they swept on after the vanishing white brush of their leader, like some great, hurrying, dark cloud-shadow, up the trail.

Anon, going always at their tireless wolf-lope that no beast in the world can outdistance in the end, they came to a village. Some of the beaters lived at this village, and had remained there. The wolves swept on and round the miserable place--some actually raced through the snowed-up street--and took up the continued trail on the other side.

Anon they came to an open plain, where the trail split, many of the beaters that were left striking away to another village where they lived; but the white wolf tore straight on along the main trail, the trail of the hunters, the attendants, and the pack-horses. And the shadows of the wolves in the moonlight kept pace with them all that terrible way.

The plain looked flat, but was gently undulating, like some gigantic ocean petrified; so that, in due time, the pack, still giving tongue wildly and terribly, saw before them, far, far ahead, a procession of dots straggling along over the endless, unbroken white. And instantly their music shut off as if at the wave of an invisible hand.

Then, as the quarry ran from scent to view, they raced. All their long, loose, nickel-steel-limbed, tireless gallop before had been nothing to their flying speed now. The taint of the blood of the slaughtered game from the chase was in their sensitive nostrils. It was like the sight of rare wines to a drunkard. Shift! Say, but the way those long-legged demons ate up the distance between them and their prey was awe-inspiring. It was uncanny. It was almost magic. It was awful.

Then things happened, as you might say, with some rapidity.

Three shots rang out in the silence--three shots in quick succession. They were fired at the wolves by the only man in the group who had an efficient rifle, but were really meant to recall the sleighs with the sportsmen *and* the rifles, which had gone on.

The wolves spread out into a long line; the ends of the line crept forward swiftly on either hand, and the whole pack raced to the attack in the formation of a Zulu impi--in the shape of a pair of horns, that is. When the points of the horns got on the far side of their "prey," they rushed together, and turned inwards, still at full gallop.

At this juncture the sleighs came back--at the gallop, too. Four .450-caliber Express rifle bullets, one .375-caliber magazine and one .315-caliber magazine bullet, arriving among the wolves in quick succession produced no confusion. Not a wolf stopped. Each beast continued its tireless gallop, swerving and dispersing as it raced, and without uttering a sound, till, almost before you could realize what had happened at all, there was a dwindling crescent of gray specks in the background, and four or five other gray shapes--two kicking--lying about in the foreground.

But--and this is where we come in--neither there in the distant snow-haze nor close in by the crowding hunting-party was the white wolf.

He had been last on view far in advance of, and heading, the point of the right-hand horn of the swiftly encircling pack--his usual place, by the way--but from the moment the returning sleighs hove in sight, and the bar-like gleam of the moonlight could be seen upon the ready rifle-barrels--he had seen that, too, and knew its meaning--he had been--nowhere.

Now, before the encircling horns of the pack closed round, one of the pack-horses, maddened with fear, had stampeded and got clear away. That horse was galloping now madly across the plain, hidden from view by a gentle swell of ground, and--the white wolf was racing alongside of it; and away behind--for few could keep up with the tremendous speed of the white wolf--another, and an ordinary gray wolf was gliding in their tracks. That was a female wolf, who more than once before had found it a profitable investment to keep her eye upon the doings of the great white leader.

She saw the white wolf leap, beheld his wrenching side-stroke at the terrified horse's throat, heard the horse scream, and watched it bound forward. Followed another leap of the relentless giant white shape; the horse seemed to stumble in full gallop, and next instant came down headlong. The rest was a whirl of snow, flying hoofs, and a horrible worrying sound. Then all settled down, and as she tore up she found the white wolf feeding ravenously against time, bolting his meal as only the wild members of the dog tribe, hyenas, and vultures can. She was starving, that she-wolf, but she halted upon her hams, such was the reputation of the white leader; but when he failed to snarl at her, she, too, fell to, and bolted her meal like a crazy thing.

Directly he had fed enough the white wolf flung round upon his heels, and, with a single quick whimper, was gone, streaking over the plain away from the hunters, away from the scattered, discomfited pack; away, away, as he had never galloped before. But, then, before he had always been the hunter. This time, if he knew anything of "Pack Law" and the temper of the pack over this bad defeat and heavy loss, coming on top of the bad bear "break"--this time, I say, it was he who was, or, at any rate, might be, the hunted. And he had reasons--very sound and private reasons--why he must not meet even one wolf of the pack in combat. Wherefore he streaked, stretched flat, and doubled into a bow, his shadow chasing him, and the she-wolf--afraid to be left alone--chasing his shadow.

Very often before the white wolf had caused the pack to run into dangerous and decimating trouble, but always with a feed at the end. He had never before sold them a pup, as the saying is, like this one. Moreover, he felt that his slaying of the horse secretly--and they were bound to scent out and read that--would not improve matters. Wherefore he guessed that, after years of restless rule, it was about time to quit, and he quitted. But unfortunately there is only one thing harder than becoming leader of a big wolf-pack, and that is, ceasing to be leader and--live.

Five miles over the desolate waste of white--and what is five miles, or ten, or fifteen, to a desperate wolf?--the two beasts ran into a river--literally into a river. Ice stretched far out from the bank, yet the river was so wide that they could scarce see the opposite bank. They could see the grinding, growling ice-blocks floating all round them which they plunged in, however, and they could feel the icy bite of the water--water that would stop the action of a man's heart.

But the white wolf did not attempt to swim to the opposite bank, or mean to. He made a detour, and landed upon the same side he had started from. He did that three times, the she-wolf always following faithfully, because she had now become too frightened to stay alone and do anything else; and then he started upon another mad gallop of miles, but this time along the bank of the great river.

Finally he stopped and stared out over the ice, the thick water, and the gnashing pack-ice. Far away, it seemed, through the snow-haze, he could see a wooded height, an immense island, round which the river looped in two great arms.

He knew the spot--trust him.

No beast in its senses would try to swim the long distance across to that island, but from time to time a hunted deer had made the attempt, and a few of those that tried it had survived the ordeal and populated the island. More than once, in heavy snow, the white wolf himself, at the head of his pack, had hunted a deer down to that very spot, and had watched its head fade across the water into the distance. Once he had started to follow; but the pack had turned back, and he at length after them, snarling at their heels. Now he knew how long the swim looked from the deer's point of view.

It was an ugly proposition. But--he turned his head in the stillness, broken only by the multitudinous voices of the ice, and heard a far, far distant multitudinous murmur, and that was no ice, and it settled him. It was the united voice of the pack *on his trail*!

He paused, ran up and down, gave an odd, little, deeply expressive whine, like a puppy afraid to take its first bath, plunged in with a rush, and struck out. Soon he was out upon a piece of drift ice, shaking himself, and began leaping from one lump of floating ice to another. It was tricky, slippery, slidy work, and a fall might mean a broken leg or a crushed skull; but anything was better than dissolving like mincemeat in the jaws of the slavering pack.

Once, when a long way out, he looked back, and beheld the she-wolf, whining piteously as if she were being thrashed--and wolves are dumb beasts when "up against it"--following him.

She, too, had heard that wild, terrifying, implacable music of the wolf-pack following them; and although I, personally, doubt if they would have touched her--unless it was the other she-wolves that did it--she seemed to have been smitten by panic, and to prefer the deep sea, or the river, rather, to that pack of maddened devils.

And so, slipping, sliding, splashing, swimming, scrambling, the white wolf, after the most appalling struggle of his life, managed somehow, blindly in the end, with sobbing breath and pounding sides, to make that terrible passage, and collapse as he landed in a stiff white heap, the water frozen in icicles upon his body as he landed.

For a long time he did not move, and it began to seem as if he had burst his heart. But at last he dragged himself to his feet and turned drunkenly--to find the she-wolf lying at his side.

Thoughts came back slowly; but at length he shook himself, and stood erect at his full, immense height. He had given the wolf-pack something tricky in the way of trails to unravel, but he knew what he had taught them too well to build too much on that. And he was right, for presently, from far, far across the water, came the unutterably terrible baying clamor of the pack, moving swiftly along--then it stopped.

For a long while he waited after that, straining ears and eyes out over the moving ice, and the water, and the night that was there; but nothing could he see, and nothing more he heard except at last, far away, one last, long, lonely, ghastlily lonesome howl--the howl of defeat.

Then it seemed--but truly it may only have been a trick of the moonlight as he snarled, revealing his white fangs under his wickedly-curled-back lip--it seemed, I say, that the White Wolf of the Frozen Waste grinned. And good reason had he to grin, for the life of the white wolf had been nothing more nor less than one long, bad, bold, blustering, bullying *bluff*! What's that? Yes, sirs--bluff! And in this wise.

Firstly, his extraordinarily long legs gave him a height out of all proportion to his real hulk; secondly, his abnormally long and woolly coat gave him an apparent bulk which was out of all proportion to fact; thirdly, his actual bulk was really scarcely larger than that of any very large wolf; and, fourthly--but this concerned him only now--he was really quite an old wolf; one long past his prime, one quite unable to face any really full-grown fine young wolf, one retaining only his matchless speed by reason of his abnormally long legs, and his leadership by terrific and cleverly acted ferocity on the strength of his apparent giant dimensions. That was all, but it was enough; wasn't it, boys? Would you care to have changed places with the old rascal, and played that bluff out against *those* odds, in *that* company, for years as he had done? I *don't* think. No, nor I, either. It was some gamble, that. What?

At last the White Wolf of the Frozen Waste turned, with an insolent flourish of his brush, and trotted up the bank on the heels of the she-wolf, who had come to life again and preceded him into the dense tangle of the woods, which swallowed him up, him and his darned bluff, utterly.

XV

FATE AND THE FEARFUL

We are the little folks--we!

Too little to love or to hate.--RUDYABD KIPLING.

No one ever accused him of not being all there. The job was to see what was there.

A tiny alderman of the red bank-vole people, whose tunnels marched with the road through the wood, taking the afternoon sun--a slanting copper net, it was--at his own front-door under the root of the Scots fir, was aware of a flicker at a hole's mouth. He looked again, and saw the mouth of that hole was empty. He blinked his star-bright eyes in his fat, furry, little square head, after the manner of one who thought he had been dreaming. But catch a bank-vole dreaming! Besides, how about the squirrel overhead? He was hanging over a branch where the flicker had been, swearing fit to slit his lungs, and old squirrel wasn't much given to make mistakes, as a rule.

The bank vole turned back into his hole, knowing the law against taking chances in the wild, and the first stride fetched him up short in violent collision with another bank-vole--otherwise red-backed field-mouse, if you like--coming the other way.

The blow, full on the forehead, did not break his neck; but it ought to have done. It cast him clean over backwards out of his own front-door, where he fell down the bank, and was received, all his little short paws scrambling for a hold, by a thistle, and would have told all the world, with a thin, high squeak, what he had sat on, if the squeak had not frozen between his chisel teeth.

There had shot out of the hole, and back, a Thing. It might have been the thick end of a whip-lash or a spring, and, like a spring, as it recoiled it coiled, and was still.

The bank-vole saw. Most entirely did he see, and felt no joy in the seeing, either. Indeed, there was no room for mistake in the zigzag black chain down the back, in the unspeakably cruel, fixed stare of the glassy, lidless eyes, in the short head and flat cranium of the true viper--viper, adder, or whatever you like to call the calamity without legs, whose other name is death.

Now, bank-voles know all about vipers. They have to; they die, else. They die anyway; but no matter, for they are small and very many. Also, vipers know all about voles, field and bank; they specialize in 'em!

But our bank-vole knew all about the "freezing" game, too, and he "froze." My word, how that little beggar was still, so utterly bereft of movement that a fly settled upon him--about the first and the last that would, I should judge! And if a learned native had come along the road at that moment--on tiptoe, of course--he would have said the viper had hypnotized friend vole with fear. Hypnotize your grandmother! But you may take it from me that serpent thing was playing his game, too. He was "freezing" to induce the quarry to move and give himself away, because, since the vole was motionless, he had no idea where the little fellow was, although he seemed to be looking straight at him--in that execrable way snakes have of seeming to look straight at everything.

You think it was a battle of patience? W-e-ll, maybe. Maybe, too, it was a battle of nerves. I like to think so, anyway, for that snake-servant of the Devil had none, and the bank-vole had; and the bank-voices broke under the awful tension--or seemed to--and the bank-vole broke the terrifying spell. Also, he broke the silence.

Away down the ditch he went, bouncing like a tiny ball of dark thistle-down, all in and out among the vegetation, which, worse luck for him, the ditch being under the accursed shadow of the firs, was scanty. And as he galloped he squeaked three times--like a little needle stabbing the late afternoon silence, it was.

His removal was one kind of quick dodge in the art of quitting; that of the viper another, and a very beastly one. The crawling thing was not much more than one-tenth of a second after the poor bank-vole in getting under way, and the rest was a--was a--oh, anything you please! I call it a sliding flicker that you rather "felt" than saw. Also, the thing rustled horribly, and Fact can say what she likes. I swear it shot along quite flat, crawling, not undulating; but, ough! what a lightning, footless, legless crawl! No wonder the poor little devil of a bank-vole squeaked! The wonder was he didn't faint on the spot, for he knew what was coming.

Up the bank he pattered, and into that, to him, great subterranean highway which seems to be conjointly kept up and used by all the mysterious little four-footed tribes of the field, and which runs the length of practically every bank and hedgerow. The place was dark and cool and echoing, and bare as the palm of your hand, and far cleaner than many palms. It might have been cleaned out that very day by a fairy vacuum-cleaner; but it hadn't. It was always like that, clean as the proverbial new pin. Heaven alone knows who did the "charing" there, but those little furry tribes might have given lessons on health in trench warfare, I reckon, at a guinea a time--and cheap at that. They had found out that dirt meant disease, you bet.

Down that tunnel drummed the bank-vole, seeking to foul his trail with just any other creature; and, the highway being, as I have said, a sort of public affair, he met first a mouse gone astray, then a mole asleep, then a long-tailed wood-mouse, then a short-tailed field-vole, then a shrew about as big as your little finger. But they must have heard the scrape of the snake's scales down that echoing tunnel following hard behind, for they avoided our bank-vole like the plague, and dived up one or other of the thousand and one side-tunnels, which opened on to the main one, too quickly for the viper to catch them.

Then the poor, little, panting bank-vole found himself once more in the open. His beady eyes shone like microscopic stars as lie paused in a copper bar of setting sunlight and looked about for a refuge. It seemed, by the piston-like throb of the whole body, that his heart would burst and slay him out of hand before the hated snake could, if he did not jolly soon find one.

Then a hedge caught his eye, and he climbed it, being a good acrobat in his spare time. Beyond, however, bringing down upon himself the pecks of several birds, he did no good, for it seemed that, whithersoever he could go, the snake could follow, and--help!--the flat, terrible head was not a yard from him now.

Worse was to follow, though. He dropped to earth again, already a beaten beast; and, to complete the catastrophe, by a miracle he had landed where there was not a mouse or mole or vole hole, or any other cover, within reach. Only one big clod of earth there was, and round that he flung himself, with that stub, scaly snout weaving at his very tail, and rolled over and over and over--done, too utterly spent even to squeak.

Then Fate lifted her finger, and things happened. All that had gone before didn't count, it seemed.

The little bank-vole was dimly aware of rolling under a big, warm, live shape. He was also aware of a funny little fussy grunt in his ear, and that a set of very white and business-like teeth flashed for an instant in the sun, as they chopped surprisedly at him going under them, and missed. Thereafter the shape sat down, nearly stifling him; and in the same instant the whole air seemed to fill with the sudden, long-drawn, venomous, terrifying hiss of the viper close at hand. Evidently the limbless death had come round the corner too quickly, and had all but rammed the shape that grunted.

I can give you my word, though, that the vole was not happy one bit. He appeared to be between the Devil and the deep sea. He had no confidence in the deep sea, or any other thing that he could think of in his world. Moreover, the deep sea, besides keeping all the air off, was most horribly bristly, even on the belly. Wherefore that vole made haste to quit station, so to speak. But in a second, it seemed, before he could clear himself, that unspeakable serpent's hiss appeared to sound in his very ear, and the deep sea, folding upon itself, made the poor vole yell as if he had touched off a live-wire. He had not, of course; but it was like being struck with a dozen pins at once. He would have got out if he could, but to move was to discover more pins, and he just had to keep where he was, squealing fit to burst.

And that saved the vole, probably.

Not that there was any magic or rubbish of that kind, of course. It was simply that the viper, shooting his every inch round the corner in the effort to grab the vole's hindlegs then or never, had hit, full pelt and nose first, the nice little array of pointed arguments carried on the back of the neck of a hedgehog, snuffing under the clod, pig-fashion, for spiders. The hedgehog, whose phlegmatic disposition and special armament allowed him the luxury of never being surprised at anything, promptly and literally shut up, so that long before the viper thing had unhooked his nose and was waving his forward part about over the hedgehog, with murder in his eye and death behind his flickering tongue, looking for a place to strike home, old hedgehog was rolled up, and snuffling and snoring away inside there, like an old man chuckling when he has just cried "Mate!" at chess.

This trying position continued for perhaps five minutes. It seemed like five days to the wretched bank-vole.

Then the slow temper of the funny old hedgepig smoldered gradually alight. His eyes grew red in the foxy head of him, his snout "worked," and he snuffled and grunted faster and faster. He made up his mind to fight. And the extraordinary combat began. Lit by the blood rays of a setting sun, from a sky all raw and red, backed by the blue-gray haze of the watching woods, the silence broken only by the ghostly whisper of the snake's scales and the tiny pig-like grunting of the hundred-spiked hedgehog, that duel started.

Peering out of a peep-hole in himself, the hedgehog waited for an opening. It was no blunderer's game, this. Death was the price of a slip. He knew, however, and accepted the risks deliberately--a plucky enough act, when you come to think of it, for a beast no more than a foot long and one and a half pounds heavy.

The opening came. Quicker than you could realize, the hedgehog half unrolled, and side-chopped with his glistening teeth. Quick, too, and quicker, the venomous, flat serpent head writhed aloft and back-lashed, swift as a released spring; but the hedgehog had ducked, or tucked if you like, more than instantly back into himself. Followed an infernal, ghastly writhing and squirming of the long, unprotected mottled serpent body as it struck--too late to stop itself--simply spines, spines only, that tore and lacerated maddeningly. Whip, whip, whip! flashed the deadly reptilian head, pecking, quicker than light flickers, at the impassive round *cheval-de-frise* that was the hedgehog, in a blind access of fury terrible to see; and each time the soft throat of the horror only tore and tore worse, in a ghastly manner, on those spines that showed no life and said no word, and defied all. It was a siege of the wild, and a terrible one.

Probably this was the first time in his life that anything had dared to stand up to that viper. He acted as if it was, anyway. Usually his malignant hiss, so full of hateful cruelty, was enough of a warning. And those who ignored that did not generally live to repeat the omission. He seemed utterly unable to understand that anything could face his fangs of concentrated death and not go out in contortions. And there were no contortions about this prickly foe, only an impassable front, or, if you love exactness, back.

Wild things, unlike man, are rarely given to lose their tempers. It isn't healthy--in the wild. But if ever a creature appeared to human eyes to do so, it was that snake. He struck and he struck and he struck, impaling himself ghastlily each time, and using up his small immediate magazineful of venom uselessly on--uncompromising spikes!

At last he drew back, a horrible affront to the fairy scene, and, in the snap of a finger, the hedgehog had unpacked himself, run forward--a funny little patter it was, much faster than you would expect--slashed with his dagger fangs, and repacked himself again in an instant.

The snake, writhing afresh under the punishment, threw himself once more upon the impassive "monkey-puzzle" on four legs, but beyond tearing himself into an even more ghastly apparition than before, he accomplished nothing. Finally he broke away, and slid off, a rustling, half-guessed, fleeting vision, and there was fear at last in those awful eyes, that could never close, as he went.

Then it was that the quiet, unobtrusive, retiring, self-effacing hedgehog threw off the mask, and hoisted his true colors. And yet, if one came to think of it, there was no cause for surprise, for was he not a member of the strange, the mysterious, the great Order of Insectivora, which includes among its members probably the most pugnacious, the most implacable, the most furiously passionate fighters in all the wild? He fairly flung himself, unrolled, and running with an absurdly clock work-toy-like gait, whose speed checked the laugh that it caused, was after that viper in considerably less than half-a-second, his eyes red as the sun they glinted in, his fangs bared for action, his swinish snout uplifted at the tip in a wicked grin. No beast to bandy words with, this. It was a fight to a finish, with no surrender save to death.

The bank-vole had already fled; but it was in the direction that the fight finally veered that he had gone, and so, peeping from between the weed-stems at the mouth of a hole, he saw all. He saw the viper, his head swaying to and fro, come sliding along, making for that very hole; he heard the sudden quick rustle in the grass behind that followed, beheld the dusky, squat form that it heralded pounce. He watched the snake's head whip round, and drive with all its power in one last desperate stroke; watched it straighten out suddenly, and recoil in an awful quivering spasm, like a severed telegraph-wire, as the hedgehog's razor-sharp teeth cut through skin and flesh and backbone; and, trembling from head to foot, he witnessed, half-fascinated, I think, the awful last threshing flurry of the viper that followed.

Later, when the moon peeped out of a hole in the clouds, and the bank-vole peeped out of one in the bank, together--and his beady eyes were not much behind the moon for brightness--when the tiny, long-eared bats were imitating black lightning overhead, and a single owl was hooting like a lost soul seeking a home, away in the black heart of the woods, the bank-vole witnessed the burial of that hated viper. It was not a big affair. Only one person--the hedgehog--took part in it, and he was singularly unhurried, for he ate that poisonous fiend all up, beginning at the tail, and thoughtfully chewing on from side to side to the head--twenty inches of snake--as if he, the hedgehog, had been inoculated in infancy, and was poison-proof.

Then, still grunting, he went away, slowly, nosing here and there, rustling loudly in that stillness, an odd, squat figure in the moonlight; and the bank-vole thought he had seen the last of him, and came out to pass about his "lawful occasions," as per custom.

Now, if you or I had taken our meals after the fashion of that "wee, timorous beastie," we should probably have departed this life from indigestion or nervous prostration inside a month.

He came very cautiously from his hole, and the first thing his fine long whiskers telegraphed him the presence of was an oak-gall--one of those round knobs that grow upon twigs like nuts, you know, but have a fat grub inside instead of a kernel. At the same instant a leaf rustled, and--flp!--there was no bank-vole.

Allowing one minute for the passing of whoever rustled that leaf, and a cloud-shadow, and there he was again, back at the gall, his shining eyes, that mirrored the moon, being the only visible part of him. He rolled the gall over and sniffed, and--that was quite enough, thank you. No nut there, and he knew it--by scent, I fancy. In that moment something trod softly, ever so softly, somewhere, and a spray of laced bracken swayed one quarter of an inch, and--the bank-vole was not.

Again about a minute's pause, and three bank-voles came out together. Our friend was the last, and another was the first, to discover a little hoard of seeds that some other tiny beastie--not a bank-vole--must have collected and forgotten all about, or been killed in the interval.

In the wild, it is the law that "they should take who have the power, and they should keep who can." It isn't a bad law, because it has much to do with that other law called the "survival of the fittest," but it is apt to come expensive if persisted in.

Our vole hopped promptly towards the other vole, and made out that the seeds were his; but before any kind of ultimatum could be delivered, a twig fell, as twigs will sometimes, for no special reason that one can see. The noise it made in that stilly wood was astonishing, and ere the twig had reached the earth there wasn't a bank-vole above ground. And yet so astonishingly quick and evasive are these little creatures that in less than thirty seconds there were the two disputants, each erect upon his haunches, with little hand-like forepaws held up and joined under the chin--as if they were actresses having their photographs taken--fighting, like little blunt-headed furies, for possession of those seeds--so it seemed. I say "so it seemed" advisedly, since close by, and almost invisible because sitting quite still, was another bank-vole, who looked as if she were waiting for something; which she probably was--a lover.

It was, however, death that came, and he is a too attentive lover. The battle had been going on some seconds without apparent result, possibly because the voles had to bite upwards, shark-fashion, owing to the fact that their fighting-teeth are wedge-shaped incisors, instead of stabbing fangs, when there was a hrrr! That is all, just like that--hrrr!

Then there were no voles; but there seemed to have been no going of the voles, either. They just were, fighting and watching the fight--then they just were not. Instead of them, on the very spot where they had been, a sheeted ghost, with wings that flapped and flapped, and never made any noise, with the face of a cat, and big round eyes that gleamed, and a snore most horrible, had simply been evolved from nowhere, and under its claws was the little red-backed lady who waited for a lover.

Now, the coming of that apparition, whose wings did not say "Hough-hough!" or "Whew-whew!" like other birds' wings do when they fly, thus proving itself, or rather herself, to be an owl, and the fight of Mr. Hedgehog and the poisoned death, had a direct connection with, and a bearing upon, the little bank-vole's life, although they may not have seemed to have at first. If the snake had not run amok against the hedgehog, the latter slow personage would have been well out in the meadow by that time, reducing the worm population, instead of hanging about and coming up the ditch at that moment, with the hot and worried air of one who is late.

What he saw was the owl on the ground, flapping her great, soft wings about, within a foot of the nicely, neatly, nattily roofed-in nest where he and his lifelong wedded wife thought they had hidden cunningly their four soft-bristled, helpless babies. What he thought he saw was the owl engaged in turning one of those same babies into nourishing infant owls' food, or "words to that effect." And the hedgehog, like most of the order Insectivora, is cursed with the temper of Eblis, too. Naturally, therefore, things happened, and happened the more hectically, perhaps, because Mrs. Hedgehog chanced at that moment to be away--attending to the last rites--shall we say?--over the form of an expiring young rat.

The little pig's eyes of him went red in his funny, bristle-crowned head, and just as a clockwork toy charges, so he charged, with a quick, grunting rustle and far greater speed than any one who knew only his usual deliberate movements would have given him credit for.

The owl had only time to turn her cat-like face and--hiss. But though that hiss would have been good enough as a bluff to frighten creatures who wouldn't upset a snake for anything, she was out of her reckoning upon this occasion. The hedgehog, who dealt in snakes as a game-warden deals in tigers, had no nerves that way. He just sailed in under the baffling, great, flapping wing, and, ere ever the bird of the night could spring aloft, had struck. It was a ghastly form of warfare, this low running in and wrenching snap. It landed right under the armpit, so to speak, and left a nasty round hole. And it is worth noting, by the way, that precisely the same sort of hole, and in the same spot almost, but lower and farther back, was to be seen upon the body of the deceased young rat that Mrs. Hedgehog was even then attending to--the trademark of the hedgehogs, that hole.

All the immediate world of the night wild, watching from grass-tuft and root and burrow, heard the rasping tap of the owl's beak hammering helplessly at the spines on the back of the hedgehog, now beside himself with rage. Not one of them, too, that did not jump with terror--engrained by the bitter experience of hundreds of generations--at her fiendish scream. Then, in a flash, that owl was upon her back, wielding hooked beak and stiletto talons, as only she knew how to use them; and the hedgehog, who had, in the blindness of his rage, run in to finish the job, shot up clean on his hindlegs, taking the clinging, flapping owl with him, while, for the first time that night, he uttered a cry other than a grunt--an odd, piercing little cry, vibrant with rage, or fear, or both. This was rather odd, because ordinarily the hedgehog is a dumb beast, who suffers "frightfulness" in grim silence.

The tables were turned now. The shoe was on the other foot, or, to be precise, the foot was on the underside. That is, the owl had got the foe where he lived, below water-line, if I may so put it, where, like a battleship, his armor did not run, and he was soft and vulnerable as any other beast. Moreover, he had not trained himself in the art of throwing himself upon his back, as the owl, who was like a cat in this particular also, had apparently done, and since he could not prance on his hindlegs, unicorn-fashion, forever, he had to come down again, belly and throat first, on that infernal battery of talons and beak.

And he got it all right enough. I give you my word that spiny one got it; but, save for that one first little cry, he took his punishment in grim and terrible silence, fighting with a blind fury that was awful to behold. What happened to him underneath there in those few brief, terrible seconds no one will ever know--and that, we may guess, is as well perhaps, for there is no sense in dwelling upon horrors. What he *did*, in the short time he was given by Fate, is a little more clear. Butting madly down, oblivious of all things, even that unspeakable fish-hook beak, grappling like a thing demented--and I think he was nearly that--he bit deep, deep down, through feathers and skin and flesh, *home*--once, twice, and again.

Then, blindly, brokenly, smothered in blood, red-visaged and horrible, he half-rolled, half blundered free of that frightful clinch, and instantly rolled up! 'Twas his habit, the one refuge of his life, so long as he breathed; his last, and usually, but not always, his first, hope.

The owl struggled somehow, in a cloud of her own feathers, to her feet. The beautiful, fan-like, exquisitely soft wings flapped and beat frantically. There came a peculiar musky sort of smell into the air. She rose, all lopsidedly, perhaps two yards, flapping, flapping, flapping with frenzied desperation, before toppling suddenly, helplessly, pathetically, as the big pinions stopped, and she collapsed sideways back to earth again, where, blood-smeared and glaring, lit by the merciless, cynical moon, she crouched and coughed--as I live, coughed and coughed and coughed, a ghastly cough like a baby's, till it seemed as if she would cough her heart up.

Then silence--that wonderful, mysterious, waiting, echoing, listening silence of the woods at night--shut down, and darkness swept over all.

When dawn came stealing westward silently over the still canopy of leaves, both combatants were still there; and they were still here, too, when the sun, silting in through a rift in the foliage, found and bathed them. The owl was crouched as she had been when the moon left her--crouched, and with her wings just a little open, like a bird about to take flight; but she had already taken wing on the longest flight of all. The hedgehog was, too, just as the moon had left him, rolled up in a spiky ball, apparently asleep; but his sleep, also, was the longest sleep of all. And over them both, in the heavy silence, could be distinctly heard that horrible "brr-brr-brr" of flies that told its own story.

Now, that was in the morning, soon after sunrise; but long before that, indeed the moment the hedgehog had first attacked the owl and forced her to turn her attention to him, the little female bank-vole, who by some mischance or miscalculation, had evaded the first terrible handshake of the owl which spells death, had rolled clear of the fight, and dashed for her life to the nearest tussock of grass that offered shelter; and the first thing she fell over there was our bank-vole, "frozen" motionless. He was there because the scene of the fight was between him and the holes in the bank, and for the life of him he could not muster up courage to run the gauntlet past those dread, struggling forms.

In the end, there being scarcely sufficient room in the tussock for both to hide effectually, and there seeming to be some danger of the combatants trampling them flat where they lay, he led the way up a tree, whose gnarled bole took the ground barely six inches away. It was one of those great-great-great-grandfather oaks, which, if it had been in a more public spot, would certainly have been raised to the dignity of one of the few hundred trees that hid Prince Charlie. It was not, however; but it had another peculiarity, as the voles found out later on.

Scared out of their little wits by the fury of their enemies below, and afraid to go down and bolt across the open, even after the cessation of hostilities, past those appallingly still, crouched bodies, who, for all they had guarantee to the contrary, might be in fiendish, alliance crouched there, waiting for them to descend, the two voles explored gradually, in their own dainty, little, deprecating, creeping way, branch after branch of the great spreading patriarch, till suddenly, at the very tip of the longest and biggest limb of all, they vanished--into ivy. What had happened was quite simple, however. There was no trick in it. It was all above-board. It was simply that the mighty tree at this spot grew close to one of those outcrops of cliff that formed, as it were, broken-off pieces from the main cliffs which bordered the river and the valley on one side farther up, and one of the oak boughs had gradually been annexed by the ivy--itself of great age--that clothed the face of the cliff.

Climbing steadily upwards through the network of ivy-stems--he had no wish to go down now, for he could hear the river talking to itself directly underneath him, and a false step meant a clean drop into the swirling black depths thirty feet or so below--the bank-vole, with his companion in close and trusting attendance, presently came out on top of the cliff. He found himself upon a space all clothed with vegetation, bushes, and stunted trees, some hundred yards long. Beneath him, as he peered over, he could see the roof of the wood, all laid out like a green tablecloth, and here and there, through gaps, the river, now shrunk to no more than a stream, by reason of the fact that men, for their own purposes, had dammed its waters about a mile farther up the valley, and constructed a reservoir there.

The voles knew nothing about any dam--then. They were satisfied to explore the cliff-top and the crevices, to discover the tiny eggs of a coal-tit, and remark on their flavor; to nose into every crook and corner that came in their way; to learn the excellent facilities the place offered for setting up housekeeping; and to discover that no other bank-voles appeared to have found their way up there.

This took time, for they naturally had to flirt in between, and so it happened that the sun had been up some while before they finally set to improvising a home, in a partially earth-filled rocky cleft, with their own sturdy forepaws. They had got so far as to dig in out of sight, turning every few seconds to push out the loose earth, when the dam up above broke, and a few hundred, or thousand, for all I know, tons of water dropped into the valley--crash!

And thus it happened that, when the sun set, those two little, big-headed, blunt-nosed bank-voles, looking out upon an endless sea of water, above which the top halves of the trees in the wood rose like mangroves, were, save for a few that had climbed into trees and would starve, the only bank-voles left alive, to repopulate that valley with bank-voles, out of all the teeming thousands whose burrows had honeycombed every bank in the vicinity. Verily, how strange is Fate, "who makes, who mars, who ends!"

XVI

THE EAGLES OF LOCH ROYAL

He makes a solitude, and calls it--peace.--BYRON.

He comes, the false disturber of my quiet.

Now, vengeance, do thy worst.--SHERIDAN.

The rising sun came striding over the edge of the world, and presented the mountain with a golden crown; later it turned the rolling, heaving mystery of the mists below into a sea of pure amber. A tiny falcon--a merlin--shot up out of the mist, hung for a moment, whilst the sun transformed his wings to purple bronze, and fell again, vanishing instantly. Next, a cock-grouse, somewhere below the amber sea, crowed aloud to proclaim the day, and a raven mocked at him hoarsely.

Then, and not till then, the Chieftain awoke. The Chieftain showed as a chocolate, golden-brown, wedge-shaped mass of feathers, perched on a lonely pinnacle of rock, and, his appalling, razor-edged claws being hidden under the overhanging feathers of his legs, he was scarcely striking. Next moment he opened his eyes, and was no longer mean, for he was a golden eagle, and the eyes of a golden eagle are terrible. In them are written hauteur, pride, and arrogant fierceness beyond anything on this earth; there is also contempt that has no expression in speech. He shot out his neck, clapped his talon-like beak, and gazed out, over the mist that hid Loch Royal, to the south shore of the loch, where lived his son. The loch was, as it were, their frontier, the boundary-line that divided the hunting-grounds of father and son, and it was seldom crossed by either bird.

A little wind rose somewhere in a mountain gorge, and went shrieking down, rending the mist asunder, as a man rends carded wool. And behind the wind slid Chieftain, who know the value of a hidden descent. He shot through the rent, racing down with the sun's rays to earth, and surprised a cock-grouse at his breakfast, nipping off the tender heather-shoots daintily one by one. So swiftly did Chieftain fall that the grouse never knew what had killed him; he was dead--in a flash. The great eagle swept on with the grouse in his claws, and, without stopping, beat upwards again.

Suddenly, without any warning, a bullet came singing over the rolling heather, and passed, with a whine, close to Chieftain's head. Later came the blasting report of a rifle. As for Chieftain, he gave one amazed scream of outraged and startled dignity, dropped his grouse, and went; and when an eagle goes in that way, it is like the passing of a rocket.

A few minutes later Chieftain was whirling round high up among the crags, calling imperiously for his wife, as a king might call. And she came, she came, that huge, fierce bird, with a trickle of blood dripping down her neck, and a fire in her eye that was unpleasant to behold. She, too, had been fired upon and grazed by a bullet, and she said so in no measured tones. Now, the laird of Loch Royal deer forests had never allowed his eagles to be fired at or killed. They were part of the family possessions, as it were--always had been for generation upon generation; and, moreover, they kept down the grouse on the deer forests--which was useful, since the grouse is the red deer's unpaid sentinel, and give him warning of the crawling, creeping stalker.

Wonderfully the two eagles circled round one another in mighty, still-winged glidings, effortless, majestic, masterly, sometimes together, sometimes apart, drawing ever away northward with scarcely a wing-flap, without, it seemed, any visible force to drive them, till they swam, like specks on the eye-ball, miles away and upwards round the white-mantled peaks.

Here, so easily can birds pass from scene to scene, they were in another world, an Arctic land, silent as the Arctic, bare as the Arctic, cold as the Arctic, and, at first sight, desolate and uninhabited as the Arctic appears to be. But this was only an example of Nature's wonderful magic. Desolate it was. Uninhabited--no.

So far as the eagles could see, there was only a raven, cursed with a far-advertising blackness, who sat upon a splintered fang of rock and mocked them hollowly. But he was not the only creature there.

Sweeping down with a hissing rush over a giddy slope of shale that looked perpetually upon the brink of a general slide down *en masse,* with their immense shadows underrunning them, the eagles startled suddenly by their unexpectedness a great red beast into motion. There was a clatter of antlers, a click of hoofs, a little shower of stones, and away went a superb stag, a "royal," a "twelve-pointer," lordly and supercilious, picking his way without a slip on that awful incline. But until he moved, even he had been quite invisible, bang in the open though he was.

The eagles, following him and swooping at him with imperious savagery, because they were still angry and upset, though never really coming near him, bustled him into taking that awful path at a loose hand canter, not so much, I think, because he, the king of the forest--and this,

this lost, lone scene, was part of the local conception of the word "forest"--cared the sweep of a "brow-tine" for the eagles, as because he was startled and uncertain as to what was supposed to be happening. And the stones spurned by his neat hoofs--he seemed to kick most of them down behind him as he finished with them, each making for itself its own miniature avalanche--helped to add to the sudden confusion.

Then it was as if a shell burst in front of him--right under his haughty nose--and he moved exactly eight feet one inch without touching the ground; also, in doing it, he cleared a five-foot-seven-inch bowlder, so absolutely without the slightest sign of an effort that he seemed to have been blown upwards. It is worth noting, because twenty seconds before he had been too lazy to clear a four-foot heather-bush, and had gone *through* it.

The "shell" had been a party of ptarmigan very much flustered and upset by being all but galloped over; not the white and frozen ptarmigan of the cheap poultry warehouse, but the "live" proposition of that name in their gray, or usual, disguise, posing as stones among many thousand that lay around the summits.

Wild horses would not have put the ptarmigan on wing in face of those terrible, sliding, underrunning shadows of death--indeed, one had been lying within two yards of the Chieftain, as he slid back low to ground after stooping at the lordly stag--but this crashing avalanche of shale with the king of the forest atop was too much for them, and they went down the "hill" into the nothing and the far distance that lay, so to speak, almost at one's feet, like a spatter of shrapnel.

At the same instant two gray shadows evolved themselves out of the very ground, and slid away, swift as scudding clouds, up the slope; and a third gray form, also apparently sprung from nowhere, rose from before them, and dropped like a spent projectile into the low-lands. They were two mountain hares and an old sinner of a gray crow; but the thing that caught the Chieftain's stabbing eye most was none of these.

Both eagles had, with half-shut wings, dropped like mighty barbs towards the dim, blue distance of the vale, after the hurtling ptarmigan; but in an instant their great vans respread, their big, wedged tails swiftly fanned, and with every available brake on, as it were, they fetched up almost short. Then they both described a single, gliding, calm, lazy-looking half-circle, and settled upon a turret rock that shot fifteen feet up from the mountain's shoulder.

Above them, the snow shimmered and glistened blindingly. Below, the warm mists of the dales steamed off under the beating sun. Loch

Royal lay like a mighty, burnished shield to the southward; and northward, peak rose behind peak in everlasting grand perspective.

Near them was only the lonely slope, bare now, it seemed, of all life. But they thought otherwise. Their unspeakably fierce gaze was focused upon what looked like a grained slab, like any other grained slab, if it had not all at once begun to twitch, and so--even then one could only make out the faint outline of a body--turn gradually into a wild cat asleep on his side. The twitching was not the result of a fit, but of dreams. Probably he had not meant to go to sleep at all--in a land of golden eagles! He had merely meant to bask in the sun, within instant spring of a handy hole between the stones if anything in the enemy line turned up. That very sun, however, had conspired with drowsiness to betray him, and--something in the enemy line *had* turned up.

Even so, I doubt if the golden eagles, with all their wonderful prismatic binocular vision, made out the cat, as man could. Birds have not that power, as man has. The twitching they were instant to see. The cause of it they must have, equally instantly, suspected. Certainly, however, was a long time coming to them. Precisely when it did, no man knoweth. They remained like carvings or very fine figures cast in bronze, and as immovable as the same, for the best part of an hour, if you please; and during that time all the sign of life that either of them gave was to wink a yellow eyelid, as quickly as an instantaneous shutter winks, several times.

At the end of that period a rain-squall came racing and howling round the summit. It passed in a few seconds, and left mist--cloud, if you like--damp, dank, and chilly, and a dead calm, in its wake, and--the Chieftain had vanished (I told you he knew the value of a hidden descent). But goodness and his own arrogant self alone knew when or where; in the squall, most likely. But he had certainly vanished.

The Chieftain's mate sat on, as stolid, and as solid, and as statuesque as ever. She had not moved when he evaporated, or given any sign whatever.

With the coming of the mist, the cat woke up. The cold probably awoke him. He was not pleased. He had come to get warm, not cold. He arose and stretched himself, baring all his claws and fangs with lazy insolence, for any whom it might concern to see.

Then--he collapsed, falling as if the slab on which he stood had slidden from under him, and remained--flattened tense, wide-eyed, and dangerous.

The Chieftain's wife had jerked her head and sneezed. At least, she had yanked her cranium quick as quick, and made a noise. It seemed like a sneeze. For the rest, she remained as motionless and expressionless as bronze Buddha, her wonderful orbs scowling at the wild cat.

Then that cat got off that slab of rock. I say got off, but it would be more correct to say that he slid off sideways on his tummy--flat. One had difficulty in seeing that he moved. His inscrutable, wide, sinister, yellow-green eyes were fixed upon the Chieftain's wife. The whole of his attention was fastened, focused, concentrated upon the Chieftain's wife. And there he made his mistake. He forgot about the Chieftain himself, but the Chieftain had not forgotten about him. In fact, the Chieftain was there, on the spot, or over it, rather, exactly above the wild cat's head, five hundred feet above, and very slowly revolving upon wide, outstretched pinions, as if hung by invisible, slowly swung wire from heaven.

If the cat had looked up! But, then, the cat would not have been a wild four-footed animal if he had. In all the aeons four-footed wild-folk never seem to have learnt to look up, and, for the omission, die some painful deaths that might otherwise be avoided. I do not say that none of them *ever* look up--they do, but it is seldom.

Indeed, finally this wild cat did look up; he could not well help it. There was a sound like the swift descent of a smiting sword above him. But he was seconds too late.

Seconds before, the Chieftain had vanished again. Nay, he had changed. His wings had shut, and--he had turned into a line, a dark streak drawn, almost as lightning draws itself, from heaven to earth, thus--wh-r-r-r-ssss-sh!

It was then that the cat looked up--just in time to receive the Chieftain's black tiger-talons, upon brilliant yellow claws, clashing against his own ivory fangs, and--well, in his eyes.

The Chieftain's wife flung from her strange self her immobility, flung out a scream, flung open her pinions, and--shifted. She could not have arrived upon the scene more than three seconds after her lord--but not by any means master. She was certainly not half that time getting to work.

I am not going to describe the struggle that followed, in deference to certain good, kind, and well-meaning people who are unable to face the stern realities of life, or--to save their country. Such things, however, must be; and they would not happen if it were not for a hard, though very sound, purpose, among beasts as among men. Nature is far-seeing and very wise. Moreover, she hates hypocrisy, and--well, we may not all be butchers, but most of us eat meat.

It was certainly a very great confloption, for, of course, that wild cat fought like a--like a wild cat, which is like a Welshman, and I cannot say more than that. And in the end the whole inferno, being upon a very sharp slope, began to slide, and slid, dragging a welter of dust and raw earth and feathers and fur after it, in an avalanche of its own, till it fetched up in a tangle of mountain-ash roots and furze two hundred feet below, where it furiously and fearfully, in one wild, awful, whirling flurry, ended.

After that the Chieftain dragged what was left of that wild cat out of the bushes, where he had tried to jamb and crawl and burrow himself, out into the open--well into the open--so that the eagles could look all round, which they like to do, being birds of high degree--also vermin, or counted as such by gamekeepers of low degree.

The pair--Heaven and the laird alone know how long they had been good and faithful partners in life--thereupon set to hooking at one another with their horny, dragon-like beaks, gripping with black-taloned yellow claws that even a Hercules would shake hands with just once, beating with monster wings that would knock you or me silly, snapping horny, resounding snaps, and generally "not 'arf a-carryin' on" in the approved and correct modern matrimonial manner. So it appeared, at least; but among eagles--within the royal circle, that is to say--such things might be their way of paying compliments, for you cannot expect feathered couples of the royal blood to behave like a pair of mere love-birds.

Then came the bullet.

It was a neat, long, nickel-jacketed, lead-nosed bullet of some .300-caliber, and its own report was chasing it. It sang a high-pitched, plaintive little song all alone to itself as it traveled along through the fine, champagne-like mountain air, at about thirteen hundred feet per second, and it was aimed to hit the Chieftain exactly in the full of the chest. That was why, I suppose, it hit the wild cat smack in the backbone, and killed that poor beast all over again. But you can never tell with bullets.

It might be mentioned here that just as turtle-soup is to their worships, so is wild cat to golden eagles--a *bonne bouche par excellence,* so to say. They do not get it every day, or every month, for the matter of that--at least, not in these islands of enlightenment, for the wild cat shares with them the honor of being a martyr of Fate, and it is on the *index expurgatorius* of the gamekeeper also.

But, I give you my word, those two mighty birds left that wild cat uneaten. I say "left" him advisedly, for it was rather a matter that they had left him than that they did leave him. Anyway, they were not near him, not anywhere near him, and I suppose they went. There had arisen a noise as if all Regent Street had at that moment rustled its combined "silk foundations," and--there were our eagles far, far away, and in opposite directions, melting quicker than real sugar-knobs in hot grog into the haze of the distant sky.

And after that the Chieftain and his wife glided up into the setting sun till it swallowed them in a red glory, and when the sun had burnt itself out, swam--swam stupendously and wonderfully--through ether down to bed. Bed with them that night consisted in sitting, regally enthroned among clouds, upon a black, rock bastion exactly above a clean drop of not much more than six hundred feet, and rocked by "the wracked wind-eddies" of the mountain-tops. The good God who made all things--even the animal that had fired at them--alone knows what they dreamt about, that superb, intolerant, fierce, haughty, implacable couple.

Now, the man, the--er--lord of creation, who had fired that shot--in fact, all those shots that day--was Pig Head, the back-to-the-lander from the South. Pig Head argued that deer forests are farms lying idle. And the laird had offered to rent him a farm at one-and-nothing-three the acre to disprove it. Pig Head had taken the offer. He disapproved of lairds as unrevolutionaries. He hated red deer because they were too smart for him to kill wholesale, and he loathed golden eagles because they were the pride of the "hills." But he kept his opinions to himself, because he valued his neck. The People of the Hills would have stretched it very much longer than his own long tongue if he hadn't. In his heart he also hated the "oppressed" People of the Hills for that they loved their laird, regarded deer-stalking as a religious rite, and--wore kilts!

As a matter of fact, Pig Head's farm never grew anything more than some clinging heather, a little cross-leaved heath, patches of furze, a clump of storm-bent Scotch firs or so, and rock--mostly rock.

Pig Head had only been able to get what he thought was his own back upon that day by firing at the eagles, because the laird and the stalkers, the gillies, the keepers, and the People of the Hills, were away, all away, at a sheep-dog trial, or a clan meeting, or something. After that he had to work in silence, and he did.

There are always people who will buy a golden eagle "British caught," and those who don't want live ones will take 'em dead, and have them stuffed. They like to be able to set 'em up in the hall among other stuffed birds, and boast that they shot 'em. Other people of it like decayed mind come and look at them, and offer money for them at sales out of jealousy. That's collecting.

Now, somewhere, somehow, sometime during his checkered career, Pig Head had heard, or read, of a way of catching golden eagles. He proceeded.

Upon an unholy and cold shaly slope well up among the clouds, the mist, and the ptarmigan, Pig Head had hollowed him out a hollow, roomy enough for himself to crouch in. He was the sort of man that crouched--and "grouched." Over the top he put a nice big slab; the walls were of piled stones, and at one end was an aperture eight inches or so long by about one foot. Being made of its surroundings, the hiding-place did not look at all suspicious--from a bird's point of view.

Finally, upon the morning after the unsuccessful shooting, and before it was light--this was necessary, for there is no knowing how far the eyes of eagles can see--Pig Head ensconced himself in this hiding-place. It was perishing cold, and Pig Head, who did not smoke, and never drank whisky--only gin--was blue of nose and numb of hand. A good plaid would have helped him, but he abhorred plaids.

The dawn came up over the mountains, the mists sank down to the vales, and the dawn wind, lean and searching, went whispering over the hills.

Then a speck grew out of the heights, out of the west and the dark, and growing and growing momentarily, became a rustling, sinister, untidy, heavy shape, which anon settled upon a rock, and croaked, "Glock! glock!" twice, almost like a bark, in a deep and sepulchral voice. To it was added another sable form, coming down from the lonely stony heights, and the two sat together, remarking, as they looked--but their wonderful eyes must have seen it very far away--at the bait. It was the wild cat turned inside-out, and other things, on a slab outside the aperture before mentioned, that was at one end of Pig Head's hiding-place. And the black specters were ravens.

"Ou!" they said; then "Aw!" then again "Ou!" One remarked "Augh!" and the other agreed--or, it may have been, disagreed--with an "Au!"

Evidently the wild cat, in a disguise in which he would not have known even his own self, looked very enticing, and he and the situation generally were being discussed from all points of view.

I say from all points of view advisedly, because, although the ravens discoursed much over their council of war, they would not come within a hundred yards, and it was a voice from the semi-dark, or western, side which finally stayed them in the very act of unfolding their big, rounded wings to fly away.

"Krar-krar-krar!" rasped the voice; and the ravens folded their wings again to wait and see!

It was a gray crow, and the ravens knew that never was gray crow an innocent lost in the wilderness.

If the gray, or hoodie, crow--always remembering that crows, gray or black, are servants of the Devil, just as ravens are, and very cunning--if the crow, I say, thought that here was food without some horrible form of hidden death lurking behind it, then the chances were the gray crow was right. They knew "hoodie," you see. Anyway, if they let him go first, and he was wrong, then it would be *his* funeral, not theirs.

Wherefore the gray crow went first to the bait, and Pig Head, half-dead with cold and peering out of a tiny peep-hole, called down blessings of a weak and watery nature upon his black head. And well he might, for if the gray crow had shied at the bait, then everybody else would have taken his tip.

They took his tip now, for in a few minutes there was a "hurrr-hrrr-hrrr" of wings, and, one after the other, down came the ravens.

Anon the ravens were joined by a third, volplaning from some cloud-covered peak, where he must have been watching all the time; and the crow was joined by four accomplices, who just drifted up from nowhere special, as gray crows have a habit of doing when there is carrion afoot.

But Pig Head had not come there to entertain ravens, nor was he at that moment laying up a store of lumbago for the purpose of gratuitously feeding disreputable gray crows. He had other quarry in view. The gray crows, however, were his best asset. They quarreled, and were loquacious, and, in fact, they made a most infernal noise; and he had stated that the noise was necessary to his success. This would seem as if eagles hunt by sound as well as by sight. Pig Head was the first person I ever heard that suggested so. But, be that as it may, the racket increased as the sun, robed in purple, gold, and crimson splendor, rose over the mountain-tops; and with the sun came the only bird, so the ancients tell us, who can look the sun full in the face without blinking--*Aquila*, the eagle.

But--make no mistake about this point--he who came then, grandly, proudly sweeping over the blue, dim ridges, was not the Chieftain himself, for this was not the Chieftain's territory, but the Chieftain's son; he who lived, as you will remember, upon the other, or south, side of Loch Royal.

Haughtily, statelily, as a king might go to his throne, so did the Chieftain's son let himself down, in stupendous hundred-foot spirals, to a pinnacle of rock, jagged, saw-edged, and perpendicular, about two hundred yards away; and the ravens and the gray crows, who saw him coming, made great and sudden hostile outcry at first, and then, as he folded, foot by foot, his immense pinions about him, and sat there erect, with his piercing, scowling gaze bent upon *them*, they were dumb.

And Pig Head, aching with cramp and cold in his hiding-place, knew that *the* quarry was at hand. But if you think, because the eagle was at hand, that the time was at hand too, you don't know eagles. They may be, upon occasion, as quick as the spring of death, but they can also be as slow as the wrath of Heaven. And that bird there, the great, grand, haughty, unbending *Aquila chrysoetus*, that golden eagle, sat. I say, he sat. And there, so far as he was concerned, appeared to be an end of it. He might have been a carven copestone of the very granite fang he sat upon, for all the appearance of life he gave, except that occasionally--say at fifteen-minute intervals--he winked a yellow-lidded wink. And the wink was almost as unlifelike and uncanny as the bird.

And the gray crows and the ravens gulped and quarreled, with one eye upon the eagle and one upon their job; and Pig Head--Pig Head sat and cursed that eagle, from his horny beak to his barred tail, through chattering--and aching--teeth. But the eagle never moved a feather.

We are told that Alexander sighed for other worlds to conquer. So it was with the Chieftain, who was not Alexander.

After his wife had gone a-hunting eastward--a wonderful and gigantic silhouette floating and dwindling into the furnace of the rising sun--the Chieftain sat upon his ledge of rock, staring across the gleaming, painted, glassy expanse of Loch Royal, southward, to the dominions of his son.

He had seen his son, a speck in the dawnlight, invisible to our eyes, sailing from peak to upflung peak. He had seen him suddenly check and circle downwards. And then--he had not seen him.

He had waited two hours, with that patience which birds and reptiles have, and still he had not seen him. Yet, if during that time he had risen, the Chieftain must have seen him. And the Chieftain knew that. He knew also that a golden eagle very rarely makes a "kill" so big that he has to remain with it two hours. The alternative, therefore, would seem to be death or carrion; and the way in which he had circled down would seem to suggest carrion. And it is written among the laws of the king of birds that when carrion is about, the strict rules and regulations as to the inviolability of the frontiers may be, in some degree, broken.

Therefore the king unfurled his overshadowing vans, and launched himself down the lake with mighty, slow, powerful strokes, like the steady thrust of marine engines. He would go and see.

Five minutes later the Chieftain was as motionless as his son, perched, like him, too, upon a rock, watching the highwaymen and footpads of the moors squabbling over the bait--they had no eyes to see what they were doing, for they had to keep one eye upon each eagle--and about two hundred yards away on the other side.

This may have hurried matters somewhat, for within only about another half-hour the Chieftain's son rose, and, with heavy wing-flaps, flew down to the bait, sending the ravens and the crows up in a cloud, like blown bits of burnt paper, as he came to anchor. And it was curious that, in stooping to meanness, the royal bird's aspect was no longer grand. He flew heavily and clumsily to the spot. He settled without grace, and almost overbalanced on to his Grecian nose. He clutched, and tore, and gulped, and gorged like a vulture. Thus Nature always dresses her actors for their parts. You may have noticed it.

"He clutched, and tore, and gulped, and gorged"—*Page 274*

[Illustration: "He clutched, and tore, and gulped, and gorged"]

But Pig Head--Pig Head was chuckling. He had silently and softly removed the clod of peat that blocked the aperture before mentioned. Running through this aperture he had a cord whose other end was fastened to the bait, and every time the great eagle wrenched and tore at the flesh, he very, very gently pulled the bait towards him. He did not move when the mighty bird had his head up, gulping, you will note; for even Pig Head knew that an eagle nearly standing on his head and tugging, and not feeling the difference between his own tugs and the tugs on a cord, is not the same as an eagle with his head up and eyes stabbing everywhere at once.

At last the victim had been drawn, upon the bait, within reach, and Pig Head, slipping his hand through the opening, grabbed the thick, powerful legs of the bird, and pulled. There was one mighty upheaval of vast vans, and--no eagle! What happened down inside the hiding-place was more or less private. There were sounds as if a young earthquake were getting ready to be born in that place; but in the end the Chieftain's son had his legs tied, and suffered the indignity of being ignominiously thrust into a filthy sack. He said nothing during that argument, but his looks were enough to kill anything with a thinner hide than Pig Head.

Immediately Pig Head got ready for the Chieftain. What's that? Yes, the Chieftain is right. That great, haughty bird had not moved. You see, eagles are not educated up to seeing their full-grown sons disappear into the bowels of the earth without explanation or warning given. There is nothing in their experience to meet the phenomenon. Consequently they don't tumble, as a rule, and--well, listen for yourself.

In a short time--a short time for an eagle; not less than half-an-hour, really--the Chieftain flapped heavily to the bait, and fed--beastily, if the truth must be told.

He was bigger than his son, and heavier, and knew more about the world, and Pig Head was longer in seeing a fair chance to make a grab at the royal legs. At last, however, the chance came, and Pig Head grabbed. The Chieftain naturally lost his balance, and before he knew what had happened he was inside Pig Head's "booby-hutch."

The Chieftain, however, was not an ordinary bird, not even an ordinary eagle. Moreover, he must have been a great age, older even than Pig Head. Be that as it may, the Chieftain believed mightily in the wild maxim which says, "They should take who have the power, and they should keep who can."

And upon that he acted.

It all happened in a flash. Like lightning his right wing came round with a terrific flail-stroke, and hit Pig Head in the face at the precise instant that the surgical instrument he carried as his beak sank deep into one of Pig Head's calves. The Chieftain was upside-down at the moment, and his legs were tied together, but that made no difference to the savagery of the blow.

Pig Head uttered one howl of agony, and tumbled backwards, and his devil saw to it that he should tumble backwards upon the very sack wherein lay the Chieftain's son, squirming with rage. The Chieftain's son was a son of his father, and hearing the young hurricane of his father's wings, and feeling the intolerable weight of Pig Head sitting involuntarily down upon him, struck for the cause like a good un--struck, with his cruel, hooked bill, through sack, through trousers, through pants, and home through flesh, and Pig Head rebounded into the air considerably quicker than he had gone down, hitting his head against the roof, a resounding whack, and yelling fit to awake all the devils in cinders. And he did not go alone. Upon one calf, and upon--another portion of him, the Chieftain and the Chieftain's son went with him.

Very few men have ever left a powder-magazine on fire in quicker time than did Pig Head leave his hiding-place, and none could have made more noise in the process. The Chieftain stuck to him lovingly, and the Chieftain's son, sack and all, seemed determined never to leave him; and Pig Head was nearly demented with pain as he leapt out, caracoling wildly, into the light of day, and into the arms of--only the laird, the head stalker, four gillies, and two collies.

They had come to find him, these stern-faced, long, lean men, on account of "information received." And they had found him. But they did not speak. They were Scotch. Nor did they screw out a smile among them. They were Jocks! They acted--being Highlanders.

Four hands like iron claws seized Pig Head, and tipped him on end, even as he had tipped the eagles. Two knives went "snick" as they opened, then "wheep-wheep" as they cut. Several pieces of cord and bits of sacking flew into the air. There was one colossal upheaval of wings, a feathered whirlwind hurling everybody every way--and the Chieftain and his son, released and scandalized, offended and enraged beyond the rage of kings, rose swiftly into the air with mighty, threshing strokes that simply hurled them aloft like powerful projectiles--into the heavens, as it were terrible avenging spirits of the tempest. A chaos, a rush, a mighty blast of air, and--they were gone!

Then the laird turned to Pig Head, and, "Mon, ye dinna ken th' laird. If ye did--w-e-e-l, Ah'm thinkin' ye'd understand."

XVII

RATEL, V.C.

Between the clumps of the stunted acacias the sun beat down with the pitilessness of a battleship's furnace, and it was not much better in the acacias themselves. Save for a lizard here and there, motionless as a bronze fibula, or a snake asleep with eyes wide open, or the flash of a "pinging" fly, all Nature seemed to have fled from that intolerable white-hot glare and gone to sleep.

But the hour of emancipation was at hand, and the dim caverns of shade--what there was of it--stirred strangely. A hundred yards away a blotch of shadow beneath a group of stunted trees swayed and broke up into several zebra moving off to water. Fifty yards distant the inky shade that carpeted the earth under a bare outcrop of rock gave up a single gnu antelope bull and a Grant's gazelle whose lyrate horns were as wonderful as his consummate grace.

Thereafter came sound. Till then there had been only heat, the first hints at movement, and the terrifying silence of the wilderness. Even the birds had been dumb. Now came "a feathered denizen of the grove" with a peculiarly arresting, grating chatter, a noise no one could overlook, and few could help investigating. And finally, brazenly, impudently, excitedly flitting from branch to branch, the chatterer evolved slowly out of the ragged bush-choked landscape, a dusky little bird, seemingly a bird of no importance, scarce larger than a lark.

Putting personal appearance aside, however, this feathered one, who dared to shatter the slumber of the everlasting wilderness, seemed to be under the impression that he was of vast importance. Moreover, his business appeared to be pressing and urgent, so that he could neither brook delay nor take "No" for an answer. It was as though he was under a desperate need to take you somewhere or show you something, and YOU must follow him--*must*; there was nothing else for it.

But nobody cared. The zebra trooped off without turning their striped heads; the gazelle, weighted under his horns, and the gnu bull stalked away unattending; the lizards remained fixed in a permanent attitude of attention; and the snakes continued to stare at nothing. No one took the slightest notice.

Then came the reply.

It was as if a person or a thing, deep down in the bowels of the earth, hearing the bird, stirred in its sleep, and shouted up, "I come." And it came.

Heralded by a peculiar, quaint, little, chatty, sibilant, hissing, whistling chuckle, there emerged from a regular cave that he, or an ant-bear, or some other burrower had constructed under an ancient bush, a beast--a most remarkable beast.

It was long--about three feet. It was low; it was stumpy, clumpy, sturdy, bear-like, and altogether odd. It had no ears that any one could find, and it rattled the most murderous armament of claws that you ever guessed at. But that was not all; not by any means. It, or, rather, he, had really been colored grayish white in the first place; but Nature had thoughtlessly dropped him into a vat of black paint on his "tummy," flat, and left him there to swim about, so that by the time he got out he was one half, including chin, black, and the other and upper half, including top of head and back and top of tail, grayish white. And then, for a joke, it seemed, Nature had painted a white band round where black and grayish white met, a sort of water-line, so to speak, and let the poor little beggar go--go, mark you, into a wild where self-advertisement is something more than unhealthful for the smaller folks. Afterwards, however, Nature--who is all a woman--had repented, seemingly, and being unable to undo her own jest, had given to the little, slow, conspicuous beast, as compensation, a courage surpassing the courage of any other beast on earth. The result was rather curious--it was also the ratel, or honey-badger, who had nothing at all to do with rats, but everything to do with honey, and was self-evidently more than three-parts badger.

"Kru-tshee! Kru-tshee-chlk! Krue-tshee-chlk-chlk, whee-tshee-tse-tse, tse-i-who-o-o!" he whistled, and chuckled, and muttered, and fairly sang to himself as he came trotting along towards the cheeky little bird, like a dog that answers a whistle. His gait was all his own, as he, too, was all his own original self, being unlike anything else, although he bore the stamp of the badger people upon him.

With a calm, rolling trot, head down, tail up, back a fraction arched, with something like the slouch of his distant relation, the wolverine, he proceeded, preceded always by that dusky phantom bird that flitted and perched ahead of him, like a yellow-hammer down a country lane--calling, calling, calling. And he, lifting his odd, flat, "earless," sleek head to it, would whistle and chuckle in reply. They had, it seemed, arrived at a perfect understanding, these two, during the centuries. "Lead on, Macduff!" he seemed to say.

They passed antelopes anchored in the shade; hartebeest, impala, and roan after their kind. They heard the click of horn and the stamp of hoof, but troubled not. They passed the place where a leopard lay asleep up a tree, and saw a devil's whip of a ten-foot mamba snake--and the bite of that same is a sixty-second short cut to the grave--flee before them as if they, and not it, were death incarnate. Once a serval cat, all legs and ears and agility, stood in their path to listen to the funny chuckling, whistling noises, but fled when it saw the little, low ratel as if it had seen a ghost.

But always undeterred by anything in the way, engrossed utterly on the task in view, the dusky bird flew ahead, calling the ratel on with its harsh cry; and always the ratel, unhurried and cool, jogged along in its wake, answering, and whistling, and chuckling away to it, as if convulsed with inward merriment. Perhaps he was. It was a strange procession, anyway, and one you don't look for every day in the week, even in Africa, the land of mysteries and surprises.

Finally, the bird stopped; and the ratel looked, and saw that it was flitting round the base of a big mimosa. Enough! He hurried a little at last. Next moment he was nearly hidden under a continuous stream of earth and dust flying back from his amazing foreclaws, and a whirling, whirring vortex of perfectly demented bees, whose nest, that had been weeks in the building, was dissolving in seconds under the trowel-like scoopings of those fearful claws.

Honey! Honey! Honey!

That was it. That was the magic word the bird, who was a honeyguide by name, had shouted to the ratel, who was a honey-badger, you remember; and honey-bees they were that made the air delirious.

The bird, with the quick eye of a detective, had located the hole of the nest, but having no trowel, forthwith fetched the ratel, who had, and together they fed, the beast on honey, and the bird on the grubs in the combs.

And the bees? Oh, they don't count! At least, they might have been house-flies for all the notice the ratel took of them, save now and then to bunch a dozen or so off his cowled head carelessly. Yet they would probably have nearly killed *us*.

It was about this time that the bull-gnu appeared, tramping steadily towards them; a rugged, rough renegade of the wilderness; a ruffian kicked--or, rather, horned--out of some herd forever, and, for his sins, doomed always to face the risks of life alone, or in the companionship of other male outlaws of soured temper like himself--almost always male; the female wild seems guiltless of law-breaking, or is under a banner of protection if it is not. Such "rogues," as men call them, are not gentlemanly, as a rule. And, by the way, you know the gnu, of course, *alias* wildebeest? The head of a very shaggy buffalo, the horsy mane, the delicate, strong, sloping antelope body, the long, mustang-like tail, and the strange, twisted, unconventional character, half-fierce, half-inquisitive.

He--that lonely one--was going to drink, and he may have been doing it early because he had only his two eyes and ears, and his one nose, to warn him of the dozen or two forms of death that awaited him at the drinking-place, instead of the eyes and ears and noses of all the herd.

The gnu saw neither honey-bird nor badger till he was within a yard of them. Then he stopped as instantly still as if he had been electrocuted.

The ratel, who had himself to feed, and a wounded wife and two young whom he would lead to that honey-feast anon, looked calmly over his shoulder at the form of the antelope towering above him. There was no sign of fear in his straight stare at the shaggy, ferocious-looking horned head. He had no business with it, and would thank it to mind its own affairs. And the honey-bird didn't care much, either, she having no young to feed, because, cuckoo-like, she left other birds--woodpeckers, for choice--to see to that.

Wherefore, for as long as a man would take to select a cigarette with care and light it, there was dead silence and stillness, broken only by the distant, deep "Hoo-hoo, hoo! hoo-hoo, hoo!" of a party of ground hornbills.

Then that devil of meddlesome curiosity which is the curse of the wildebeests fell upon that gnu, and sanity left him.

"Kwank!" neighed he. And again, "Kwank!"

Next instant he had spun, top-fashion, on all four feet at once, and jumped in the same manner, and was gone, whirling round them, with great shaggy head down, and in a halo of his own swishing tail, at the rate of knots.

It was nothing to be wondered at in that strange antelope that he should then sink from wild motion to absolute, fixed rigidity, broken only by the restless, horse-like swishing of the long tail, staring hard at the ratel.

Perhaps it was the bees that did it, or perhaps the ratel stood in the gnu's very own path, or in the way of his private dusting-hole. I know not; neither did the ratel--nor care much, for the matter of that. But when the gnu went off again, circling with hoarse snorts, and shying and swerving furiously and wonderfully at top speed, he sat up on his hindlegs, the better to get a view of the strange sight. Perhaps he thought a lion was lying somewhere near that he could not see from his lowly, natural position.

Again the gnu stopped as utterly instantly as if he had run into a brick wall, pawed, stamped, snorted, and went off once more into furiously insane caperings--a new set--all the time circling, with the little, black-and-gray, erect figure of the surprised ratel as a pivot.

And then, in a flash, before any one had a second's warning to grasp the truth or prepare, with head down, eyes burning in the down-dropped, shaggy head, and upcurved horn-points gleaming in the afternoon sun, he charged, hurling himself, a living, reckless, furious battering-ram, straight at the little ratel.

Did that ratel quit quick? Do ratels ever quit an unbeaten foe? I don't know. They may, once in the proverbial blue moon; but I haven't seen 'em. This one didn't. He seemed to know that it is held to be a sound military maxim to meet an attack by counter-attack, and he did, though he had only the fifth of a second to do it in. Ah, but it was good to see that odd little beast trotting out coolly, head low, tail high, singing his war-song as he rolled along to meet the charging foe so many, many times his own size.

Next moment there was a thud--somewhat as if some one had punched a pillow--and the ratel was flying through the air, high and fine, in a graceful and generous curve. A thorn-bush--what matter the precise name? there are so many in those parts, all execrable--acknowledged receipt of his carcass with a crash, and for a few seconds he hung, like a sack on a nail, spitted cleanly by at least one thorn, far thornier than anything we know here, before the thing gave way, and he fell, still limply, this way and that, hesitatingly, as it were, as each point lovingly sought to retain him, to a fork near the bottom, where he stayed.

At last he picked himself out of the fork, and--oh my!--with a whistling grunt of rage, coolly, calmly, clumsily if you like, but grandly all the same, trotted forth into the open to look for that bull-gnu again. And that, sirs, was the sort, of animal *he* was.

The bull-gnu, however, who was not previously acquainted with small beasts that would face his charge--and an aerial journey, *and* the thorns--and come back for more, had fetched a curve at full gallop, and loped off into the landscape. For the first time since the herds outlawed him, I fancy, he seemed to be quite pleased with himself, and soon, antelope-like, put the ratel from him placidly, and forgot. But he was reckoning without his host. If he had done with the ratel, the ratel had not done with him. No, by thunder--not by a good bit!

Finding no bull-gnu, the slow little black and grayish-white fighter from Fightersville returned at a walk, still whistling with rage, to the unearthed bees'-nest, which looked like a town after a bad air-raid. And the first thing he did was to patter almost on top of a cobra, a five-footer, who, having narrowly escaped death by the gnu's flying hoofs, was what one might call considerably "het" up, or "off the handle," so to say.

The servant of the Devil sat up, blew out its beastly hood, and shot forth a hiss that seemed to run all up and down one's spine, like lightning on an elm-tree.

The robber of honey sat up, said "Tchik!" and turned a somersault. What's that? Yes, somersault is right.

Followed instantly two thin jets of liquid, as much as anything I can think of like those lines called "trajectory curves" which ballisticians do so love to draw in books on rifle-shooting; only, these curved lines began at the hollow point of Mr. Cobra's poison-fangs, and were meant to end in Mr. Ratel's eyes. They didn't. Old man ratel, he was standing on his hindlegs, with his sturdy paws in front of his eyes--like a man who looks across a sunny land--and seemed just about to turn a somersault again. He changed his mind, though, when the poison, that would have blinded him for life--and that life wouldn't have been long in that wild *then*, I want to tell you--stopped, and he went in at that black-necked, legless, soulless servant of Satan, utterly and amazingly unafraid. It was fine.

Oh, by the way, I forgot to tell you that when Nature repented, and gave the ratel a courage surpassing the courage of any other beast on earth, she also gave him a skin tough as a pachyderm's, and loose, as if it were two sizes too large; and that is why that black-necked cobra died quite quickly, and the ratel didn't, even slowly. Even if the snake's fangs had got through, which was not in the least likely, that did not mean to say they would touch Mr. Ratel's person inside. This, by the way, may explain why being spitted on thorns, like a beetle on a pin, when the bull-gnu charged, did not seem to worry him much, either.

The moon was up when the wounded mother ratel, on guard at the mouth of her burrow, looked up sharply. A side-striped jackal, who kidded himself she had not seen him lying in wait to find out, when she went hunting, what she hid in that den, suddenly bolted with a yap; and a hyena, represented by two burning eyes, who appeared, by some magic of his own, to guess she was wounded, jumped up and made way for something that approached. It was her husband and the cobra, the latter trailing along limply behind, who came that way; and even the hyena had retired, with an audible sigh--at least, it wasn't a moan quite--when he claimed the path. After all, there is no sense, if you are the most cowardly beast for your power on earth, in getting up against the pluckiest thing in creation in full possession of life and liberty.

Later our ratel sallied forth to "face the world" again. His wife had recovered from her wounds--the result, these, of refusing to believe she was not so good as a twelve-foot python, and a bit better--sufficiently to walk slowly; but that was not enough to face that wild where die-quicks, from lions, down through leopards, hyenas, wild-dogs, jackals, and the rest, are forever hiding, on the lookout for unfortunate ones flying an S.O.S. signal. No, he must go and do the provisioning alone, and alone he went.

For a peaceful beast, one only too pleased to mind his own business and thank other folks to mind theirs, his subsequent doings were rather astonishing. This was because he cared for neither man nor beast nor devil, in the first place, and because the night produced all three, in the second.

He got man in the form of the smell of meat--well-seasoned meat, even for Africa!--what time he was testing a native village, by scent and on the downwind side of him--and that showed his pluck, my word!--for honey or fowl. He detected neither out of the few dozen unspeakable stenches, but struck meat, and following it up-wind, arrived at a piece--a good big piece--on the ground among grass.

A civet cat--who is more civet than cat, by the way--a small spotted genet--who looked like an exaggerated ferret in the uncertain gloom--and the inevitable black-backed jackal--who must not be confused with him of the side-stripes--faded out at his approach like steam in a dry atmosphere. He might have felt proud of this silent respect, if it were not a fact that these gentry, these village frontier haunters, scenting danger, thought it a fine "kink" to let the brave one test it first.

And he did.

To be exact, that ratel touched off the tooth-jawed trap that was the reason for that free meal of high and valuable meat in that place, and when he jumped he didn't get anywhere. Also, it hurt his leg abominably.

Then the others reincarnated themselves out of the shadows--especially the jackal, who shouted "Yaaaa-ya-ya-ya!" and called a friend--

and waited for things to happen. They were confident things would happen, for Africa is not a good place wherein to get caught in a trap--*there is too much likelihood of being mistaken for the bait*!

But they might as well have seen a thunder "portent" captured by the tail as this ratel by the leg; for, instead of instantly and foolishly abandoning himself to the frenzy of unthinkable fear--the fear of being trapped is the greatest of all to a free, wild thing--as practically all others would have done, he said nothing at all; he failed to lose his head; and, to crown all, he instantly, coolly, slowly, viciously, and doggedly set himself to struggle, with a grim persistence that was amazing. And, moreover, from that instant he never left off.

A striped hyena, seemingly in lifelong terror of his own shadow, turned up by magic--or perhaps he heard the snap of the trap. Seven times he bolted, for no earthly reason that one could see, before finally gaining courage to snap at the ratel at the very end of his reach. It was the kind of snap that would take half a man's face away, and not nice to meet when you are trapped. The ratel, however, came calmly at the hyena, trap and all, and so nearly got his own trap-jaws locked home on the unclean one that the hyena was glad to go away.

In the end, thanks to the amazing toughness of his skin, and its looseness, the ratel managed to, as it were, slide the bone of his leg between the jaws of the trap, leaving the skin and fur in, and the rest was mainly determined tugging and strong fang-work.

Then he coolly ate the real bait, and--the onlookers remembered appointments elsewhere. None of them, it seemed, was tickled to meet the ratel when he had finished. He was sure to be crusty; and, anyway, he had bitterly disappointed them all--he had achieved the apparently impossible, and, worst part of the lot, was not dead.

Now, a ratel will do almost as much for honey as a bear for pork, a leopard for little "bow-wows," or a man for diamonds. This will explain why he was foolish enough to follow, some hours later, the trail of some natives who had been out collecting honey from a camp the day before; or perhaps he knew nothing about the honey till, not too scientifically, he got into the camp. Anyway, the honey was very good.

There are, however, from a wilding's point of view, camps and camps. Most of the inhabitants of the wild, including the lion, who are not born with a pluck considerably above proof, can discriminate the difference. The ratel either could not or would not.

Then the knowledge was driven home. Driven home in the shape of a big, loose-limbed, deep-jowled brute of a dog, as unlike the ordinary native curs as it well could be. It did not come silently, or suddenly, for it growled full warning in a terrible bass; but the ratel showed contempt, and teeth that glistened beautifully in the red light of the dying fire the sleeping sentry ought to have seen to, but had not. Moreover, it did not come alone, for the camp was a white hunter's camp. The dog gave a thunderous baying rally-call, and almost before that sentry had leapt to his feet, the ratel vanished tumultuously and suddenly from the public gaze, under a perfect cloud of dogs. He was, ere any one knew what the riot might be, literally smothered under dogs--dogs, too, most of 'em who held up the deadly leopard, and hounded the tyrannical lion, habitually and for a pastime, mark you.

Then his devil prompted one of the black sentries to rush up and fire his rifle. Probably he did not know what was under those dogs; certainly he thought it would keep there. In any case, he nearly killed a dog, and the cause of the trouble did not keep there. He came out, miraculously alive, still more miraculously cool and unhurried. He broke away from the dogs as if they were little puppies, and, still quite coolly and slowly, he charged that man.

The yell that followed could have been heard quite a long distance through the cloaked night. And, in truth, one cannot wonder, for you may take it from me that the jaws of a ratel fast home on the calf of your leg, as our ratel's jaws were on that native's leg, form something to remember in dreams.

But it was that very native who saved our ratel's life, all the same; for his gymnastic display during the few seconds that followed was so energetic that the pink pyjamas and a revolver, that represented the white hunter fresh from sleep, had no chance at all of doing any damage except to the dancing native--which they nearly did; and the dogs, once more piling themselves on to the ratel, broke his hold, and the whole fight rolled and raged away into the darkness and the thorn-scrub, out of sight.

Later, one by one, those dogs came back, dead-beat most of them, with tongues lolling and sides pumping. Some limped, and some turned away every few yards feverishly to lick a wound. All were blood-stained, but not a drop of it--not one drop--belonged to friend ratel. He, that superb warrior, was at that moment trotting along, quite unconcernedly, through the bush about a quarter of a mile away. There was blood upon him, too--not his, the dogs'--and no other mark; and though he was pretty sore and sick from internal bruising, his skin, his wonderful loose skin, was whole, and unpierced by a single fang. He had, however, the decency to go home and fling himself into a stupor-like sleep, just to prove that he was a real, live beast of this earth, and not merely a phantom from other worlds.

The next afternoon was closing in dull and cloudy, and there were signs of a dark and bad night to come--just the sort of day wild hunters come out early in. This was why the grunt sounded then that heralded the appearance of our ratel above-ground, and he himself appeared, emerging at his very own slow trot from his hole. For a moment he paused, looking round, with his funny, "earless," flat head in the air, as if he expected, or listened for, the honeyguide; but the honeyguide was half a mile away, leading some natives--who, by the way, were endeavoring to copy the crooning, whistling replies of a ratel--to honey.

No honeyguide? Then he must go and search for himself. And he did, returning, in fifty minutes, for his wife, who, now much recovered--as only a ratel can recover from the very jaws of death--followed him with her young to the hole he had torn in a rotten tree-trunk where the bees were nesting.

They had proceeded perhaps three hundred yards, when, turning a bush carelessly, as no other creature would dare to do, the ratel fell almost on to the back of the bull-gnu.

There is no need to be surprised that they should meet. The wild is not an aimless mix-up in that way. Each creature has its beat, temporarily or permanently, nor seeks to deviate. You may look for the same herd of antelope, feeding near the same place, about the same hour each day; the same lion stumping the same beat, as regular as a policeman, most nights; the same hyena uttering horrible nothings within hearing of the same hills, any time after the setting of each sun, just as surely as the same cock-robin asks you for crumbs, the same blackbird awakens you with inimitable fluting, and the same black cat seeks for both in the same vicinity each dusk.

The surprise was in what followed. Perhaps the bull-gnu kicked our ratel badly as he lurched to his feet, jerked from half-sleep into violent collision with he knew not what. Perhaps the ratel had a memory. Perhaps the presence of his family weighed with him. Whatever the cause, the result was decided enough. He reared and hit deep, and fixed home a very living vise, where he bit.

Then things happened, but that which immediately followed was not a fight; it was not even a spar. The ratel never moved, although he was moved--astoundingly. The gnu bull did the moving, and produced the most amazing bit of violent activity one could dream of. It was quite indescribable. A buck-jumping mustang of the most hustling kind would have been as a gentle lamb to it. The ground all about looked as if herds had jumped upon it--bushes, grass, flowers, and all were trampled down flat. But it did not do what it was designed to do--it did not break the ratel's hold. Bruised, assuredly, shaken so that he ought to have fallen to hits, dizzy and blind, he did not let go, and in the position he held he could not be hammered off. He just glued where he was, saying nothing at all, till the end--till that grand old bull sank and was still, exhausted, by loss of blood, and with one great hopeless sigh his life departed from him, and he died.

The ratel did not leave go for some little time. He seemed to suspect that the gnu bull was bluffing, or perhaps he was himself half-stunned.

It was the sudden and peculiar growling hiss from his wife--sounding all a-magnified in that wilderness silence after the battle--that made him look up, at her first, and then almost instantly at *something else*. His wife was backing slowly towards the "bush," every hair on her body sticking straight out at right angles, her eyes fixed strangely upon that something else. His young had taken to cover, not, it seemed, too readily, but by their parent's order.

A lion was standing, still as a carved beast, at the far end of that little clearing--*he* was the something else. Goodness and his kingly self alone knew how long he had been there, that great, heavy-jowled, deep-bellied, haughty-eyed brute. He may have been present from the first, or the middle, or only at that moment. Being a lion, he was just there, suddenly, without any visible effect of having got there, a presence of dread, created apparently out of thin air at the moment, in that spot, and with less sound than a blown leaf.

This power of being, without seeming to come, of evolving from nowhere, is one of the lion's most highly perfected tricks; for King Leo believes in all the ritual of his craft, and is great on effects, even down to the minor details. Power, grim and terrible, he has, without shadow of doubt; but he never forgets to impress that fact--and more--upon the world, and every action is carefully studied to advertise, not himself, but his "frightfulness." A very fine play-actor is the king of all the beasts.

But the ratel did not move. He had met his Napoleon, and was not--so far as the watcher could see--afraid.

Motionless, scowling, with head down, and shrewd, proud eyes smoldering, the lion stood there like an apparition of doom. He was, I fully suspect, letting the effect sink in deliberately. He knew his game. Also, he had a reason. Surely a great poker-player was lost in the lion.

But the little ratel met that regal stare squarely and unmoved. He whose proud boast it was that he feared nothing that walked or crawled, or swam or flew, could not be frightened now. And he who came to terrify was perhaps all but ten feet long, and he whom he sought to terrify was barely three feet. It was a comparison to make you gasp.

Now, that lion did not want the ratel at all, or his wife, or his family, or anything that was his. He wanted the gnu, and would be very pleased if the ratel would go away and leave it to him.

The ratel, moreover, did not want the gnu, being an eater of honey, locusts, and generally badger-like fare for the most part; and if the lion had only had the sense to wait a few minutes longer behind the scenes, the ratel would have gone away and left the gnu. But he would not be driven; *that* was the rub. Attacking nobody unprovoked, he was a grim beast to attack, and gave way before none. Hence the trouble.

Finding that the bluff of the impressive tableau did not work, the lion tried a fresh one. Still staring at the ratel, he sank his head to the ground, so that his great mane hung to the earth all about him. His forelegs and his shoulders crouched, but his hindlegs and his back were held at their highest, and his tail began to lash behind. Then he began to growl tremendously and nerve-shatteringly, and as he did so he curled his upper lips up and back, till the whole ghastly array of his teeth was laid bare to view. In this position he looked like a gigantic grinning mask, with blind eye-sockets where the wrinkles were on the sloping forehead, his eyes nearly invisible below, and a tail lashing far up atop. It was a horrible sight, and one calculated to stampede the pluckiest animal.

It was, of course, also a deliberate piece of mesmeric bluff, the reason for which was not made clear till one noticed, what the ratel probably could not, that the great leonine tusks, the terrible fangs, were yellow and worn, as were the rest of the teeth. This was an old lion, a king on a throne already tottering, a monarch of yesterday.

That lion, however, might have turned into Satan himself, for all the ratel cared. He was threatened, attacked, bullied, forced. His blood was up, and had not all who ever fought him allowed that he was the pluckiest beast on earth? Enough! Come lion! come devil! he would give ground to none.

"All allowed that he was the pluckiest beast on earth"—*Page 296*

[Illustration: "All allowed that he was the pluckiest beast on earth"]

Lions are not too patient. Also, they have fine spirit of their own. They are among the very few beasts who will hunt and attack animals as strong as, or stronger than, themselves. And this lion's patience snapped suddenly. All at once he seemed to remember that he was still a king, though a king already within the shadow of abdication. The terrible bass rumble of his growl grew, and changed tone; his tail lashed faster and faster; and then, all suddenly heralded by a couple of wicked, rasping, coughing grunts, he--charged.

The ratel moved to meet him--to meet him--and at a cool jog-trot!

What happened then was hard to follow. It looked as if the worn fangs of the lion failed to make his hold on the wonderful, leathern, loose armor of the little honey-badger, and that he bungled the stroke of his terrible paw. Be that as it may, the honey-badger certainly went straight in, right under the lion's guard, right under the lion, and rearing, he bit home, and hung like a living spanner.

And here, perhaps, it is best to draw a curtain. For one reason, I cannot describe it, and frankly confess the fact. For several other reasons, it is best not to try. The ratel died in about ten minutes, crushed, battered, smashed to death; but the chaos lasted longer than that, because, even after death, he was not done with--the passing of life had locked his amazing jaws shut forever, and *they were shut on the lion.*

The end found the little ratel lying crumpled up and crimson on the trampled grass, and the lion running about like some great injured dog, squatting down every few seconds to lick furiously at his wound. Fear was in the eyes of the king of beasts, for the first, probably, and certainly for the last, time in his life, and his blood reddened the grass wherever he made his way; but the internal hemorrhage was the worst.

Then the vultures came, and that, my friends, is a signal for us humans to go. The vultures get the last word always, even in a story, and the name of that word is--FINIS.

XVIII

THE DAY

Now, if you wore a helmet and neck armor of purple, green, and blue in metallic reflections, with scarlet cheek and eye pieces, if your uniform were of purple, brown, yellow, orange-red, green, and black, "either positive or reflected," with a long, rakish, dashing rapier-scabbard cocked jauntily out behind, wouldn't *you* feel proud? So did he; pride and the "grand air" were written all over him. True, though, the rapier-scabbard was not a rapier-scabbard exactly--only a tail; but it looked like one, in a way. His full title was *Phasianus colchicus,* but ordinary people called him just plain pheasant for short.

You would have thought, after all this, that even in the first pale light of a cold dawn he would have been easy to see. As a matter of fact, Gaiters, the head gamekeeper, one of his underlings, three dogs, and a gun passed right under his bed without seeing him. Rather, they may have unconsciously seen him, and put him down as a bundle of dead twigs and leaves caught up in a branch. This is not very complimentary, perhaps, to a gentleman attired in a gorgeous uniform as heretofore set out, but true; and lucky for him, too, to have at once a uniform of unquestioned splendor and one which would melt into its surroundings.

They, the men, did not see him; but he, the old cock-pheasant, saw them right enough. He opened one eye, and stared at them through that. Then he opened the other eye, and stared at them through that. Neither stare seemed to please him.

It was not Gaiters's way to march through the wood at that hour in the morning. What meant this unseemly disturbance of *Phasianus's* domain? His suspicions, never long at rest, woke up. Moreover, somewhere at the back of his brain rose a memory, a little, tiny speck of a memory, which grew.

Then he stood right up.

Everywhere, here and there, in the gray cloak of the dawn-mist, he could hear the sharp "Chawk-chawk!" and the quick, flustered whir as pheasant after pheasant came down from its roosting-perch to clean and breakfast. But his suspicions held him for a few minutes longer, stretching his neck and peering about at the still-shrouded mystery of the ground below; and it was as well, perhaps.

Suddenly his head was still; a spray of a brier-bush was swaying gently. There was nothing in that, of course, if there had been any breath of wind to move it; but there wasn't. Wherefore our gallant friend did not come down to preen and breakfast like the rest, but sat motionless as a statue, while the sun rose and touched him to a winking golden and bronze wonder, and the mists began to be torn asunder. He wanted to know what moved that brier-branch, and he wasn't taking chances till he did.

Day came on apace, and all the night hunters who had remained so late had already hurried off to bed save one. *He* appeared, evidently empty, certainly very angry indeed at having waited for a cock-pheasant who refused to do what he was supposed to do and come down to breakfast. Out of the brier-bush he came, a lean dog-fox, snarling horribly up at the pheasant, who calmly returned the gaze, conscious of his safety, of course, and said "Chuck it!" in a loud, harsh voice, and quite distinctly, twice.

The fox, knowing it was no good to wait any longer in the daylight, went, like a floating red shadow. The pheasant watched him go, but did not move for some time. Foxes had been known to come back again more suddenly than they went.

At last he flew to the earth, but even then he did so as silently as his noisy wings would let him; and he did not announce the fact with a half-crow, as the others had done. Very circumspectly he slipped off through the undergrowth, by a series of little crouching runs, stopping every now and then to freeze and listen.

Soon he came to one of those open, beautiful, grass-covered "rides" with which keepers intersect pheasant-coverts. He stopped dead on the edge of it, himself invisible among the drooping, leaning, old-gold bracken. The "ride" was full of wood-people, for here had been scattered that corn which Gaiters intended the pheasants to feed upon. Indeed, there were about ten pheasants, hens and young cocks of the year, doing exactly what they were intended to do. Also, there were some half-dozen softly-tinged, blue-gray wood-pigeons, and one cheeky jay--whose wing-patches rivaled the perfection of the blue sky above--doing their best in a quiet sort of way to help the pheasants, which they were not intended to do, by any manner of means.

The old cock-pheasant slid across the "ride" after a bit, low as a crouching rat. He had no business there this day. His mind was still alert with suspicion. Moreover, his father had been a cunning old cock who had managed, by ways that were dark, to keep out of the game-bag for years, too. The taint, as Gaiters would have called it, had been passed on to him.

He made for the open edge of the covert, and he was mighty careful about doing that even. He felt that air and plenty of horizon were necessary to his well-being, after the disturbing vision of Gaiters and Co., so unnaturally busy, hurrying through the dawn.

Now, it is quite remarkable how much you can see from the edge of a pheasant-covert without being seen yourself. Keepers know that, but do not give the fact away. The ground sloped away in two open grass fields, a hedge dividing them, and it was within about the longitude and the latitude of where that hedge met the covert that our old friend maneuvered.

The climate about there seemed to suit him admirably. True, good food was not strewn in plenty just where he could most easily see it. He had to look for his acorns or his beechmast by the good old domestic-fowl plan of scratching among the leaves; roots also he was forced to scratch for; and the noisy mistle-thrushes with the tempers of Eblis had to be driven off the berries he would look after in the ditch.

Also, there was a stoat. That stoat, however, tackled him just once. In the process it discovered (*a*) that he wore spurs not meant to be ornaments, and (*b*) that no one could teach him much about using those same spurs. The stoat, plus a new carmine decoration for gallantry, remembered an urgent appointment down a rat-hole, and kept it. Perhaps it was a young stoat, and had not learnt that there are at least four degrees of cock-pheasant, namely, young and brainless, adult and brave, old and brave and cunning, and old and decrepit; but the last stage is a rare bird. There is nothing of any use to the stoat in the second and third degrees of cock-pheasant--no health to the stoat, you understand.

The dawning of the morning had passed by now in gold and crimson and purple splendor; the mist-curtain had been drawn back by the fingers of the wind, the utter darkness upon everything at ground-level had begun to give way before the sun, and to leeward of most trees and bushes there was a balmy luxuriance of golden light that held one lingering.

Gnats were dancing under the low-hung boughs in still corners, as they will dance on the coldest day; song-thrushes were beginning to take life for one more day, and tack hither and yon, as if they were busy pegging the field down with an invisible "cats' cradle"; and the black rooks, shining like burnished steel shields, flashed and flashed again as they began to gather beneath the trees, where the ground thawed most and first. Though they alone seemed to have discovered it, the pigeons very quickly found that the rooks had hit something good, and you can bet one jay, at any rate, must be there, to make profit out of somebody.

It was the jester of a jay, whom, in spite of his painted plumage, no one seemed to have noticed, that first gave the alarm that carried the cock-pheasant's suspicious temperament a step farther upon the path of independent action. Up till then you will note that, though he had left the vicinity of his own people, he had not yet left the realm that was peculiarly their own--the woods.

"W-a--r-k, w-a-a-r-k!" gasped out the jay suddenly, and fled, a half-seen vision of pinkish, of black and white, upon uncertain, almost fluttering, wings.

It was like striking a gong. Instantly all motion was suspended, and dotted thrushes, clustered rooks, and deprecating pigeons remained at rigid attention. The old cock-pheasant, too, erect as an armored warrior, unseen just within the covert, stood promptly at gaze.

Then, in no more time than one would take to inhale one puff from a cigarette, the fields were empty--stark, cold, and deserted in the eye of the morning sun. The birds had not so much gone, exactly, as simply faded out--dissolved, as a picture may at a cinema-show. The cock-pheasant did not go. He was in cover, and had a good view, a strategic position of some moment.

Followed a pause. Then a man in tweeds entered one of the fields by the gate. Followed him two more, then a fourth, then two not in tweeds, then dogs, black and big, to the number of three, not to mention the bar-like gleam given off by the barrels of the guns that the first four carried. The whole procession passed silently, as they thought--but to the waiting, watching, wild-folk unpardonably noisily--diagonally across the field, and out of sight round a bend of the wood. They had an air about them. I don't know what it was exactly, but you could feel they were going to do something serious that had not been done there for a long time. Perhaps the old cock-pheasant felt it too, but--well, there now! Where had the old "varmint" gone?

Half-way down the hedge, very low and long, the cock-pheasant was sneaking. He seemed suddenly anxious to mind his own business, and that everybody else should mind theirs. He was going away from the wood, which the books tell us is the realm, the sanctuary, the all, to a pheasant, and he had no desire to answer questions by the way. For this reason, then, and a few others, he felt no special delight in sighting, about two hundred yards farther on--at a place where two stacks surrounded by rails stood and sheltered a fowlhouse--a baker's dozen of fowls sunning themselves on the hedge-bank. He held for fowls all the wild creatures' contempt for the tame or domestic. All the same, he saw no health in risking the open just then, and would not turn back, so there was nothing for it but the fowls.

Low as low he crouched, and ran very quickly, and hoped for the best; and there is no bird that can wish itself out of sight in this fashion better than friend pheasant. But he forgot the odd cockerel out. He shot right on to the wretched thing--a gawky red youth--messing about all alone in a nettle-clump, and it dashed into the field, racing on long yellow legs, and squawking fit to wake the dead.

Down clapped the pheasant as if the noise had pierced his heart, and remained stiller than the crawling roots around him, and not half so easy to see. But it was no good. Up shot the dozen heads above the herbage, and two dozen vacuous eyes regarded his vicinity with empty-headed inquisitiveness.

He almost melted into the ground, but it was useless. An old, old hen--who perhaps was ignored by the lord of the harem, and hoped for an adventure--waddled up, stood within a yard of his crouched, rounded shape without seeing him, saw him, shot straight up in the air at least one foot, screaming for help, and promptly charged blindly into the hedge, where she as promptly got held up among roots and twigs.

The old pheasant got to his feet just as the rooster who owned the outfit came racing up, panting and red. He had heard a wife scream for help. Perhaps it was the odd bird out; or, anyway, some one who had to be abolished. And he never waited to think. He saw what might have been a small cockerel (if it had been large he might have thought twice) crouching, and--he just sailed right in.

Then something happened. The two met, going up breast to breast. For a moment or two the cock-pheasant showed on or about that big rooster. Some feathers hung in the air. The rooster sat on his heels, met by a blow in the chest that seemed to take all the wind from his sails, so to speak, and would have drawn off to reconsider things if he had not promptly become more busy than ever before in his life.

It was over ere any one knew quite what was happening. The old cock-pheasant had passed through the crowd and vanished at the double down the hedge, and the big rooster was slowly subsiding into a pool of his own blood, from which he was destined never to rise again.

But those who make, instead of following, their own destiny do not get let off thus lightly in the wild. The pheasant had not gone a hundred yards, when a most intolerable blast, an almost unbearable blast, of shrill, nerve-racking noise throbbed through his head. The bird fell in his tracks where he ran, as if some one had jerked his legs from under him, and he peered out.

What he beheld was an under-keeper standing close by and blowing upon a two-note pea-whistle till there seemed some danger that he would burst his cheeks, or a blood-vessel, on the spot, and far up the field three wandering pheasants racing back to the covert, as they thought, for very life; but, as a matter of fact--and you shall see--it was to very death. The blower of whistles was stationed there to drive back into the covert any pheasants who were so misguided as to wish to roam thence into the fields and away.

Now, that old reprobate of a pheasant of ours was a pretty confirmed runner, anyway. He had trained himself to it. Yet never in all his checkered life was he conscious of a more awful desire to flee by means of the wings that God had given him. The weakness was over in a few seconds, and he crept on; but it was a near thing while it lasted. He passed, however, away from the danger zone, resisting temptation, and it was as well.

As he went there burst forth, at the opposite end of the big covert to that at which he had come out, a sudden, quick shot. It echoed away and away back among the woods, clattering and banging like great doors shutting. The old cock-pheasant stopped to listen; he cocked his green head on one side; he stood with one foot daintily uplifted: and in the same instant there burst upon the air a rending, crashing succession of shots, worse than ragged volley-firing, which almost made him jump. It had begun--the big shoot over *his* covert, the largest, the best, the richest in pheasants, which had been saved for this--"the day" had begun. When it ended very few pheasants would be left alive, for word had gone forth that it was to be thinned down, almost shot out, and that not a cock must escape.

He, our own cock-pheasant, might have chuckled--as a cock-pheasant can, and will, very low and softly to himself, if you are close enough to hear him--if something had not very suddenly and very mysteriously said "phtt!" just like that, close beside him. The old bird's head snicked round, right round, almost hindpart before; but he made no other movement. The sound was new to him, and of a strangely sinister import. Also, there was a little splayed hole in the ground, as if a walking-stick had been poked in there, close beside him, which had not been there before.

He was still staring when something, singing a little, high-pitched song in a minor key to itself, came romping through the silent air, and, with an oddly emphasized and emphatic "phtt!" landed between his feet. It bored a hole just like the first thing, and it spat dirt up into his face.

The third mystery thing clipped three feathers from his back as he ran, bolted for dear life, crouching low--even then he would not rise--for the hedge. He got there alive, if not quite whole; while a fourth nameless object cut twigs off above him. Then he kept on running, always hugging the hedges, till he was two fields away. He was upset and overstrained, for Fate had given him plenty of deaths to circumvent as it was, in the ordinary course of business, and this addition was a bit too much.

There are other forms of shooting pheasants than the orthodox one, which begins with smoking a cigarette on a comfortable shooting-seat, and ends with a wild and furious fusillade, using three guns as fast as you can. So thought the farmer's son, who took the chance to test his new American .22-bore repeating-rifle, now that all the keepers were well out of the way. And he had come mighty close to bagging the old cock-bird, too. "As near as made no odds," he said, which was true, but only the old bird himself knew quite the closeness of the call.

In the far field the bronze king of the woods found peace for a bit. The stunning reports in the covert not far away, and the thought that his companions of yesterday, his lady-loves of last spring, were even then being butchered by the hundred, made no difference to his digestion. He fed on with that imperturbability that must have come to him straight through his ancestors from the East--Kismet! It was sufficient.

He ought, of course, to have been in the covert. He was, however, here--knowingly here, cunningly here, safely--No, by Jove! Not that, by any means.

A head, clean and neat and sharp, had poked out of the long, pale grass at the edge of the hedge-ditch, and stared at him. He couldn't very well miss seeing it because of the unforgettable brightness of its beady eyes, and the absolutely spotless purity of its white shirt-front. Besides, he knew the owner--and its reputation.

He was helping the farmer to clear an oat-stubble of charlock-seeds at the moment, and bending down. That is to say, he was doing inestimable good, for which he got no credit. The next moment, and the next, and for many more, he was still bending down. In fact, from the instant he got sight of that head, it was as if a Hand had come down and turned him by magic into a big model of a bird cast in bronze. All life in him appeared to have dried up and fled. He looked as if you could have picked him up and put him upon a bracket in your drawing-room without his ever moving again. But that was only because of the head he had seen--and its reputation. Moreover, the head was not alone. At least, it had multiplied itself half-a-dozen times in less than half-a-dozen seconds, and even a stoat, which the head belonged to, cannot be in two places at once--though for sheer quickness of movement it, and far more its cousin the weasel, comes very near to it.

Just at this moment it seemed that about the roost unhealthful thing he could do would be to be seen in the air. Wherefore did this innocent and guileless old bird affect not to see the stoats, but made out that he was feeding his way along, quite and absolutely intent upon that yellow devil of a weed whose other name is charlock. He did not even hurry, and each deliberate step was taken with almost a proud daintiness. The only thing was, he never lifted his head; he was almost too obviously unwary--for him. And he gave the impression that every step would be his last out into the field; that he was always going to turn back next instant or the next, as he had done before when the stoats were not there.

On and on he kept till he had crossed the field, going faster and faster, till he ended at the far hedge with a run. And there, so far as he was concerned, was an end of the stoats. He put them aside. He forgot all about them.

They, however, had not forgotten about him.

It was half-an-hour later, and he was patiently gleaning such food as the rooks and the sparrows and the larks had left behind them, when something, he could not tell what, caused him to straighten up, with that beautiful, proud bearing that seems part of the pheasant's heritage from the gorgeous East.

And he was only just in time.

The stoat that had come up behind him, unseen, turned on its heels as it charged, changing its mind at the last moment, as if it saw he saw, and was gone again before you could click a finger, diving superbly back into long grass.

They were following him, then, those little hounds of death; tracking him; running him down. And why? He did not know, perhaps, yet--maybe he did. Blood is a dangerous thing to have on you in the wild--a flaming signal of distress for eye and nose to detect--and they are not often rescuers who hurry to the scene. He had blood on his back, that cock-pheasant, and just every now and then a single bright drop fell by the way. The .22-bore bullet had only grazed him. 'Twas nothing--but it bled more than you would expect. And that explained it. The tracking stoats thought he was wounded.

But even then the old cock-pheasant would not rise.

The firing in the covert had risen suddenly to a fierce crescendo, breaking out afresh from another quarter. Here, however, was silence--the absolute, deadly silence in which all the weasel tribe hunt. But they were there, though he could not see them. He knew that, invisible even in the sunlight--they were closing in, tracking him fast, those stoats.

Then he ran. He ran not so much for his life, but for the right to keep on the ground. If the worst came to the worst, he could always fly; but he would do anything rather than that.

He turned and ran away from the woods--raced like a fowl, but quicker, lower, much harder to see. A sudden gleam of bright-chestnut fur dead ahead, however, stopped him, and he turned back, keeping always to the hedge--towards the covert. He could hear no sound around him, only the burst and the bang of the guns in the woods, and he might have been alone; but directly he came to another hedge, and swung down to it at right angles, a furry tail with a paint-brush tip, vanishing round a holly-stem, fetching him up all standing. They were there, too, those stoats. He seemed to be surrounded on all sides save one, and that the one towards the woods. So he swung back into his original path.

Then, very soon, as he ran up the hedge-ditch, it seemed to him as if the dead leaves collected there were beginning to whisper behind him. But there was no wind to move them. Moreover, it grew closer, till it seemed at his very tail, that whisper of dead leaves.

Then, in a flash, he had stopped, spun about like a top, and struck with his spurs twice--whack, whack--more than instantly, and a long, low, brown body--close behind him, that had risen as he turned, so that its spotlessly clean shirt-front offered him a fine mark, went over sideways--with a grunt and all the wind knocked out of it, as well as an inch-and-a-half gash to remember friend pheasant by. That was one stoat; but it was not alone. He had a vision of chestnut forms sliding and rippling in and out of the shadows and the long copper gleams of the westering sun.

As he turned again, and the whispering began once more behind, the firing in front broke out afresh, and much nearer. Still he would not rise, however. It was this fact, probably, which kept the stoat-pack at his heels. They seemed convinced that he was badly wounded and unable to fly.

Then came the road. He was on it before he knew. There was the wedge-shaped, low-browed head of a stoat racing up along one side of him, with murder plainly written in the gleam of its beady eyes; there was the hot breath of another beating on his opposite flank; there was one with feet out and all brakes on, trying its best to pull out one of the feathers of his long and beautiful tail; and--there was the road dead ahead.

It was one of three--the road, the air, or death where he was. He chose the road, and crossed, like a hunted cat crossing a back-yard. His feet seemed scarcely to touch the dust as he negotiated the open, yet he had time to take in a fact or two. One was that the stoats had stopped--a little bunch of peering heads on a group of craning necks on the edge of the ditch behind him. Another was that several people and a motor-car were standing still in the road quite close, watching the shooting. I don't think any of them saw him, but he felt as if all of them did.

Arrived in the hedge on the far side of the road, he clapped down, panting. The hedge ran along the road. On the other side of it was the grass of the park-land, stretching away two hundred yards or so to the edge of the covert, which came down to a point here. He could hear the tapping of sticks in the covert--beaters' sticks. He could hear an occasional shout. Men in tweeds stood motionless on the edge of the covert, and suddenly moved.

Then came the infernal crash of the guns again, and he saw a hen-pheasant pitch sickeningly on her head from a height, and a cock-pheasant, flaming like a rocket in the sinking sun, run the gauntlet of four shots, only to turn over and slide down at a fifth.

Then--and then, he jumped.

Something had pushed past him. In the din he had not heard it. He turned as he crouched, and saw that it was a hen-pheasant, with blood on her breast and one wing trailing alongside. And in the same instant he was aware of a man--an under-keeper--crackling about in the hedge only ten yards away, looking for that hen-pheasant.

And the unwounded old cock, crouching almost till he looked like a tortoise, followed the blundering, staggering, wounded hen. It was the only thing he dared do.

It was a strange creep, and an erratic one, with many stops, those two hunted ones took together, meeting, so strangely, too--not for the first time, since she had been one of his wives in the dim peaceful past--with the guns thundering away so close, and their sons and their daughters being slain almost all around them. They had, however, little time to think about it, for they came, after about twenty yards, to a gap spanned by barbed wire, and they stopped, the cock about a foot behind the hen's tail, in cover scarcely enough to hide them.

But that was not all. Two men in fawn overcoats stood in the road by the gap, looking through it at the shooting; and a boy with a bicycle stood close to them, interested in the same thing.

It was the boy with the bicycle that did it; or, rather, it was the unhappy hen-pheasant that made him. She, being *in extremis,* had made some noise among the stiff dead leaves. It was not much of a noise, but it caught the boy's young ear, and he bent forward to peer at the hedge.

One of the men saw him, said something, to which the boy nodded, jumped down into the ditch, and thrusting in a long arm, began to feel with a purposeful hand. The hen-pheasant, whose nerves were already shattered to little pieces, struggled to get out of reach, and in a second had given the whole show away.

But I like to think of what our cunning old cock-pheasant did then. He did nothing--absolutely nothing at all. Crouching as flat as an overturned saucer, just, behind the hen-pheasant's tail, he remained stiller than a bunch of dead leaves, and far more silent. And this, mark you, when the hen-pheasant was pulled out, frantically fluttering and helpless, and there and then had her neck wrung in front of his very eyes. That, my masters, needed a nerve, after all that he had gone through. What?

The two men, seeming to think that they had got enough for one quiet walk, departed, not quickly, but without unnecessary delay. The man who had been looking for the hen-pheasant, and had seen nothing of what took place at the gap, gave it up, and went away over the grass to the shooters. The shooting ended with one last double shot, at one last old cock-pheasant driven reluctantly from the last hush of the covert; the dogs were out, galloping all over the ground for the wounded and the slain; the watchers in the road departed; the shooters gradually merged into groups, and drew farther and farther away up the park; and the boy, who was shy, mounted his bicycle and rode off into the sad blue-gray of the gathering dusk.

The big day was over, and the old cock-pheasant was alone with the melancholy song of a single robin, and a chaffinch calling "Chink!" And the cold breath of the sunset wind, shuddering and sighing all to itself across the face of the empty scene, touched the feathers that were left by the hen-pheasant attached to thorns and twigs in her last struggle, so that they danced and wavered and flickered before the old cock's eyes, as a reminder of all that had been for them in the past--the past, which for him, but never for her, might be again.

That night he roosted in the covert, as usual.

THE END

* * * *

Made in United States
Cleveland, OH
12 December 2025

28227028R00132